WORLD HEALTH ORGANIZATION

INTERNATIONAL AGENCY FOR RESEARCH ON CANCER

IARC MONOGRAPHS

ON THE

EVALUATION OF THE CARCINOGENIC RISK OF CHEMICALS TO HUMANS

Polynuclear Aromatic Compounds,
Part 3,
Industrial Exposures in Aluminium Production,
Coal Gasification, Coke Production,
and Iron and Steel Founding

VOLUME 34

This publication represents the views and expert opinions
of an IARC Working Group on the
Evaluation of the Carcinogenic Risk of Chemicals to Humans
which met in Lyon,

18-25 October 1983

June 1984

INTERNATIONAL AGENCY FOR RESEARCH ON CANCER

IARC MONOGRAPHS

In 1969, the International Agency for Research on Cancer (IARC) initiated a programme on the evaluation of the carcinogenic risk of chemicals to humans involving the production of critically evaluated monographs on individual chemicals. In 1980, the programme was expanded to include the evaluation of the carcinogenic risk associated with employment in specific occupations.

The objective of the programme is to elaborate and publish in the form of monographs critical reviews of data on carcinogenicity for chemicals and complex mixtures to which humans are known to be exposed, and on specific occupational exposures, to evaluate these data in terms of human risk with the help of international working groups of experts in chemical carcinogenesis and related fields, and to indicate where additional research efforts are needed.

This project was supported by PHS Grant No. 1 UO1 CA33193-02 awarded by the US National Cancer Institute, Department of Health and Human Services.

ISBN 92 832 1234 7 (soft-cover edition)

ISBN 92 832 1534 6 (hard-cover edition)

PRINTED IN FRANCE

CONTENTS

NOTE TO THE READER

The term 'carcinogenic risk' in the *IARC Monographs* series is taken to mean the probability that exposure to the chemical will lead to cancer in humans.

Inclusion of a chemical in the monographs does not imply that it is a carcinogen, only that the published data have been examined. Equally, the fact that a chemical has not yet been evaluated in a monograph does not mean that it is not carcinogenic.

Anyone who is aware of published data that may alter the evaluation of the carcinogenic risk of a chemical to humans is encouraged to make this information available to the Unit of Carcinogen Identification and Evaluation, International Agency for Research on Cancer, 150 cours Albert Thomas, 69372 Lyon Cedex 08, France, in order that the chemical may be considered for re-evaluation by a future Working Group.

Although every effort is made to prepare the monographs as accurately as possible, mistakes may occur. Readers are requested to communicate any errors to the Unit of Carcinogen Identification and Evaluation, so that corrections can be reported in future volumes.

NOTE TO THE READER

[illegible]

[illegible]

[illegible]

[illegible]

IARC WORKING GROUP ON THE EVALUATION OF THE CARCINOGENIC RISK OF CHEMICALS TO HUMANS: POLYNUCLEAR AROMATIC COMPOUNDS, PART 3, INDUSTRIAL EXPOSURES IN ALUMINIUM PRODUCTION, COAL GASIFICATION, COKE PRODUCTION, AND IRON AND STEEL FOUNDING

Lyon, 18-25 October 1983

Members[1]

P.A. Bertazzi, Institute of Occupational Health, 'Luigi Devoto' Clinica del Lavoro, University of Milan, via S. Barnaba 8, 20122 Milan, Italy

E. Bingham, Vice-President and University Dean for Graduate Studies and Research, Professor of Environmental Health, University of Cincinnati, Mail Location 627, Cincinnati, OH 45267, USA

L. Fishbein, Acting Deputy Director, National Center for Toxicological Research, Jefferson, AR 72079, USA

G.W. Gibbs, Director, Health and Safety Affairs, Celanese Canada Inc, PO Box 6170, Station A, Montreal, Quebec H3C 3K8, Canada

P. Grasso, Robens Institute of Industrial and Environmental Health and Safety, University of Surrey, Guildford, Surrey GU2 5XH, UK

J.M. Harrington, Institute of Occupational Health, The Medical School, University of Birmingham, University Road West, PO Box 363, Birmingham BI5 2TT, UK

P. Hitcho, Assistant Director, Safety and Health Department, United Steelworkers of America, Five Gateway Center, Pittsburgh, PA 15222, USA

J. Lewtas, Genetic Bioassay Branch (MD-68), US Environmental Protection Agency, Research Triangle Park, NC 27711, USA

G. Matanoski, Department of Epidemiology, School of Hygiene and Public Health, Johns Hopkins University, 615 N. Wolfe, Baltimore, MD 21205, USA (*Chairperson*)

---

[1]Unable to attend: T. Buckley, East Midlands Gas Board, PO Box 145, De Montfort Street, Leicester LE1 9DB, UK; C. Redmond, Professor of Biostatistics, Graduate School of Public Health, University of Pittsburgh, 318 Parran Hall, Pittsburgh, PA 15261, USA

S. Mazumdar, Department of Biostatistics, Graduate School of Public Health, University of Pittsburgh, 318 Parran Hall, Pittsburgh, PA 15261, USA

J. Melius, Chief, Hazard Evaluations and Technical Assistance Branch, National Institute for Occupational Safety and Health, Robert A. Taft Laboratories, 4676 Columbia Parkway, Cincinnati, OH 45226, USA

J.E.H. Milne, Chief Occupational Health Officer, Health Commission of Victoria, Public Health Division, Occupational Health Service, 5 Macarthur Street, Melbourne, Victoria 3002, Australia (*Vice-Chairperson*)

T. Sanner, Chief, Laboratory for Environmental and Occupational Cancer, Norsk Hydro's Institute for Cancer Research (The Norwegian Radium Hospital), Montebello, Oslo 3, Norway

A. Tossavainen, Department of Industrial Hygiene and Toxicology, Institute of Occupational Health, Haartmaninkatu 1, 00290 Helsinki 29, Finland

W.E. Wallace, Emerging Energy Technologies Team Leader, Division of Respiratory Disease Studies, National Institute for Occupational Safety and Health, 944 Chestnut Ridge Road, Morgantown, WV 26505, USA

Representative of the National Cancer Institute

J.A. Cooper, Acting Associate Director, Carcinogenesis Extramural Program, Division of Cancer Cause and Prevention, National Cancer Institute, Landow Building, Room 8C41, Bethesda, MD 20205, USA

Observers[1]

British Cast Iron Association and British Foundry Association

P.F. Ambidge, BCIRA, Alvechurch, Birmingham B48 7QB, UK

Commission of the European Communities

M.-T. van der Venne, Health and Safety Directorate, Bâtiment Jean Monnet, Plateau de Kirchberg, BP 1907, 2920 Luxembourg

International Primary Aluminium Institute

P.J. Lawther, c/o International Primary Aluminium Institute, New Zealand House (9th floor), Hallmarket, London SW1 4TQ, UK

[1]Unable to attend: J.H. Stenmark, American Iron and Steel Institute, 1000 16th Street NW, Washington DC 20036, USA; T. Thorslund, RD-689, Chief Biostatician, Carcinogen Assessment Group, US Environmental Protection Agency, Washington DC 30460, USA; J.M. Launder, 15 Church Close, Horsell, Woking, Surrey GU21 4QZ, UK (European Coking Industry)

Secretariat

H. Bartsch, Division of Environmental Carcinogenesis
J.R.P. Cabral, Division of Environmental Carcinogenesis
M. Friesen, Division of Environmental Carcinogenesis
L. Haroun, Division of Environmental Carcinogenesis (*Co-Secretary*)
E. Heseltine, Research Training and Liaison
J. Kaldor, Division of Epidemiology and Biostatistics
A. Likhachev, Division of Environmental Carcinogenesis
D. Mietton, Division of Environmental Carcinogenesis
R. Montesano, Division of Environmental Carcinogenesis
I. O'Neill, Division of Environmental Carcinogenesis
C. Partensky, Division of Environmental Carcinogenesis
I. Peterschmitt, Division of Environmental Carcinogenesis, Geneva, Switzerland
S. Poole, Birmingham, UK
R. Saracci, Division of Epidemiology and Biostatistics
L. Simonato, Division of Epidemiology and Biostatistics
L. Tomatis, Director
H. Vainio, Division of Environmental Carcinogenesis (*Head of the Programme*)
J. Wahrendorf, Division of Epidemiology and Biostatistics
J. Wilbourn, Division of Environmental Carcinogenesis (*Co-Secretary*)
H. Yamasaki, Division of Environmental Carcinogenesis

Secretarial assistance

S. Cotterell
M.-J. Ghess
M. Lézère
S. Reynaud

IARC MONOGRAPH PROGRAMME ON THE EVALUATION OF THE CARCINOGENIC RISK OF CHEMICALS TO HUMANS[1]

PREAMBLE

1. BACKGROUND

In 1969, the International Agency for Research on Cancer (IARC) initiated a programme to evaluate the carcinogenic risk of chemicals to humans and to produce monographs on individual chemicals. Following the recommendations of an ad-hoc Working Group, which met in Lyon in 1979 to prepare criteria to select chemicals for *IARC Monographs* (1), the *Monographs* programme was expanded to include consideration of exposures to complex mixtures which occur, for example, in many occupations.

The criteria established in 1971 to evaluate carcinogenic risk to humans were adopted by all the working groups whose deliberations resulted in the first 16 volumes of the *IARC Monographs* series. This preamble reflects subsequent re-evaluation of those criteria by working groups which met in 1977(2), 1978(3), 1982(4) and 1983(5).

2. OBJECTIVE AND SCOPE

The objective of the programme is to elaborate and publish in the form of monographs critical reviews of data on carcinogenicity for chemicals, groups of chemicals and industrial processes to which humans are known to be exposed, to evaluate the data in terms of human risk with the help of international working groups of experts, and to indicate where additional research efforts are needed. These evaluations are intended to assist national and international authorities in formulating decisions concerning preventive measures. No recommendation is given concerning legislation, since this depends on risk-benefit evaluations, which seem best made by individual governments and/or other international agencies.

The *IARC Monographs* are recognized as an authoritative source of information on the carcinogenicity of environmental and other chemicals. A users' survey, made in 1984, indicated that the monographs are consulted by various agencies in 45 countries. As of June 1984, 34 volumes of the *Monographs* had been published or were in press. Four supplements have been published: two summaries of evaluations of chemicals associated with human cancer, an evaluation of screening assays for carcinogens, and a cross index of synonyms and trade names of chemicals evaluated in the series (6).

---

[1]This project is supported by PHS Grant No. 1 U01 CA33193-02 awarded by the US National Cancer Institute, Department of Health and Human Services.

3. SELECTION OF CHEMICALS AND COMPLEX EXPOSURES FOR MONOGRAPHS

The chemicals (natural and synthetic, including those which occur as mixtures and in manufacturing processes) and complex exposures are selected for evaluation on the basis of two main criteria: (a) there is evidence of human exposure, and (b) there is some experimental evidence of carcinogenicity and/or there is some evidence or suspicion of a risk to humans. In certain instances, chemical analogues are also considered. The scientific literature is surveyed for published data relevant to the *Monographs* programme; and the IARC *Survey of Chemicals Being Tested for Carcinogenicity* (7) often indicates those chemicals that may be scheduled for future meetings.

As new data on chemicals for which monographs have already been prepared become available, re-evaluations are made at subsequent meetings, and revised monographs are published.

4. WORKING PROCEDURES

Approximately one year in advance of a meeting of a working group, a list of the substances or complex exposures to be considered is prepared by IARC staff in consultation with other experts. Subsequently, all relevant biological data are collected by IARC; recognized sources of information on chemical carcinogenesis and systems such as CANCERLINE, MEDLINE and TOXLINE are used in conjunction with US Public Health Service Publication No. 149(8). The major collection of data and the preparation of first drafts for the sections on chemical and physical properties, on production and use, on occurrence, and on analysis are carried out by SRI International, Menlo Park, CA, USA, under a separate contract with the US National Cancer Institute. Most of the data so obtained refer to the USA and Japan; IARC supplements this information with that from other sources in Europe. Representatives from industrial associations may assist in the preparation of sections describing industrial processes. Bibliographical sources for data on mutagenicity and teratogenicity are the Environmental Mutagen Information Center and the Environmental Teratology Information Center, both located at the Oak Ridge National Laboratory, TN, USA.

Six months before the meeting, reprints of articles containing relevant biological data are sent to an expert(s), or are used by IARC staff, to prepare first drafts of monographs. These drafts are then compiled by IARC staff and sent, prior to the meeting, to all participants of the Working Group for their comments.

The Working Group then meets in Lyon for seven to eight days to discuss and finalize the texts of the monographs and to formulate the evaluations. After the meeting, the master copy of each monograph is verified by consulting the original literature, then edited by a professional editor and prepared for reproduction. The aim is to publish monographs within nine months of the Working Group meeting. Each volume of monographs is printed in 4000 copies for distribution to governments, regulatory agencies and interested scientists. The monographs are also available *via* the WHO Distribution and Sales Service.

5. DATA FOR EVALUATIONS

With regard to biological data, only reports that have been published or accepted for publication are reviewed by the working groups, although a few exceptions have been made: in certain instances, reports from government agencies that have undergone peer review and are widely available are considered. The monographs do not cite all of the literature on a particular chemical or complex exposure: only those data considered by the Working Group to be relevant to the evaluation of carcinogenic risk to humans are included.

Anyone who is aware of data that have been published or are in press which are relevant to the evaluations of the carcinogenic risk to humans of chemicals or complex exposures for which monographs have appeared is asked to make them available to the Unit of Carcinogen Identification and Evaluation, Division of Environmental Carcinogenesis, International Agency for Research on Cancer, Lyon, France.

6. THE WORKING GROUP

The tasks of the Working Group are five-fold: (a) to ascertain that all data have been collected; (b) to select the data relevant for evaluation; (c) to ensure that the summaries of the data enable the reader to follow the reasoning of the Working Group; (d) to judge the significance of the results of experimental and epidemiological studies; and (e) to make an evaluation of the carcinogenicity of the chemical or complex exposure.

Working Group participants who contributed to the consideration and evaluation of chemicals or complex exposures within a particular volume are listed, with their addresses, at the beginning of each publication. Each member serves as an individual scientist and not as a representative of any organization or government. In addition, observers are often invited from national and international agencies and industrial associations.

7. GENERAL PRINCIPLES APPLIED BY THE WORKING GROUP IN EVALUATING CARCINOGENIC RISK OF CHEMICALS OR COMPLEX MIXTURES

The widely accepted meaning of the term 'chemical carcinogenesis', and that used in these monographs, is the induction by chemicals (or complex mixtures of chemicals) of neoplasms that are not usually observed, the earlier induction of neoplasms that are commonly observed, and/or the induction of more neoplasms than are usually found - although fundamentally different mechanisms may be involved in these three situations. Etymologically, the term 'carcinogenesis' means the induction of cancer, that is, of malignant neoplasms; however, the commonly accepted meaning is the induction of various types of neoplasms or of a combination of malignant and benign tumours. In the monographs, the words 'tumour' and 'neoplasm' are used interchangeably. (In the scientific literature, the terms 'tumorigen', 'oncogen' and 'blastomogen' have all been used synonymously with 'carcinogen', although occasionally 'tumorigen' has been used specifically to denote a substance that induces benign tumours.)

(a) Experimental Evidence

(i) *Evidence for carcinogenicity in experimental animals*

The Working Group considers various aspects of the experimental evidence reported in the literature and formulates an evaluation of that evidence.

Qualitative aspects: Both the interpretation and evaluation of a particular study as well as the overall assessment of the carcinogenic activity of a chemical (or complex mixture) involve several considerations of qualitative importance, including: (a) the experimental parameters under which the chemical was tested, including route of administration and exposure, species, strain, sex, age, etc.; (b) the consistency with which the chemical has been shown to be carcinogenic, e.g., in how many species and at which target organ(s); (c) the spectrum of neoplastic response, from benign neoplasm to multiple malignant tumours; (d) the stage of tumour formation in which a chemical may be involved: some chemicals act as complete carcinogens and have initiating and promoting activity, while others may have promoting activity only; and (e) the possible role of modifying factors.

There are problems not only of differential survival but of differential toxicity, which may be manifested by unequal growth and weight gain in treated and control animals. These complexities are also considered in the interpretation of data.

Many chemicals induce both benign and malignant tumours. Among chemicals that have been studied extensively, there are few instances in which the neoplasms induced are only benign. Benign tumours may represent a stage in the evolution of a malignant neoplasm or they may be 'end-points' that do not readily undergo transition to malignancy. If a substance is found to induce only benign tumours in experimental animals, it should nevertheless be suspected of being a carcinogen, and it requires further investigation.

Hormonal carcinogenesis: Hormonal carcinogenesis presents certain distinctive features: the chemicals involved occur both endogenously and exogenously; in many instances, long exposure is required; and tumours occur in the target tissue in association with a stimulation of non-neoplastic growth, although in some cases hormones promote the proliferation of tumour cells in a target organ. For hormones that occur in excessive amounts, for hormone-mimetic agents and for agents that cause hyperactivity or imbalance in the endocrine system, evaluative methods comparable with those used to identify chemical carcinogens may be required; particular emphasis must be laid on quantitative aspects and duration of exposure. Some chemical carcinogens have significant side effects on the endocrine system, which may also result in hormonal carcinogenesis. Synthetic hormones and anti-hormones can be expected to possess other pharmacological and toxicological actions in addition to those on the endocrine system, and in this respect they must be treated like any other chemical with regard to intrinsic carcinogenic potential.

Complex mixtures: There is an increasing amount of data from long-term carcinogenicity studies on complex mixtures and on crude materials obtained by sampling in an occupational environment. The representativity of such samples must be considered carefully.

Quantitative aspects: Dose-response studies are important in the evaluation of carcinogenesis: the confidence with which a carcinogenic effect can be established is strengthened by the observation of an increasing incidence of neoplasms with increasing exposure.

The assessment of carcinogenicity in animals is frequently complicated by recognized differences among the test animals (species, strain, sex, age) and route and schedule of administration; often, the target organs at which a cancer occurs and its histological type may vary with these parameters. Nevertheless, indices of carcinogenic potency in particular experimental systems (for instance, the dose-rate required under continuous exposure to halve the probability of the animals remaining tumourless (9)) have been formulated in the hope that, at least among categories of fairly similar agents, such indices may be of some predictive value in other species, including humans.

Chemical carcinogens share many common biological properties, which include metabolism to reactive (electrophilic (10-11)) intermediates capable of interacting with DNA. However, they may differ widely in the dose required to produce a given level of tumour induction. The reason for this variation in dose-response is not understood, but it may be due to differences in metabolic activation and detoxification processes, in different DNA repair capacities among various organs and species or to the operation of qualitatively distinct mechanisms.

Statistical analysis of animal studies: It is possible that an animal may die prematurely from unrelated causes, so that tumours that would have arisen had the animal lived longer may not be observed; this possibility must be allowed for. Various analytical techniques have been developed which use the assumption of independence of competing risks to allow for the effects of intercurrent mortality on the final numbers of tumour-bearing animals in particular treatment groups.

For externally visible tumours and for neoplasms that cause death, methods such as Kaplan-Meier (i.e., 'life-table', 'product-limit' or 'actuarial') estimates (9), with associated significance tests (12,13), have been recommended. For internal neoplasms that are discovered 'incidentally' (12) at autopsy but that did not cause the death of the host, different estimates (14) and significance tests (12,13) may be necessary for the unbiased study of the numbers of tumour-bearing animals.

The design and statistical analysis of long-term carcinogenicity experiments were reviewed in Supplement 2 to the *Monographs* series (15). That review outlined the way in which the context of observation of a given tumour (fatal or incidental) could be included in an analysis yielding a single combined result. This method requires information on time to death for each animal and is therefore comparable to only a limited extent with analyses which include global proportions of tumour-bearing animals.

Evaluation of carcinogenicity studies in experimental animals: The evidence of carcinogenicity in experimental animals is assessed by the Working Group and judged to fall into one of four groups, defined as follows:

(1) *Sufficient evidence* of carcinogenicity is provided when there is an increased incidence of malignant tumours: (a) in multiple species or strains; or (b) in multiple experiments (preferably with different routes of administration or using different dose levels); or (c) to an unusual degree with regard to incidence, site or type of tumour, or age at onset. Additional evidence may be provided by data on dose-response effects.

(2) *Limited evidence* of carcinogenicity is available when the data suggest a carcinogenic effect but are limited because: (a) the studies involve a single species, strain or experiment; or (b) the experiments are restricted by inadequate dosage levels, inadequate duration of exposure to the agent, inadequate period of follow-up, poor survival, too few animals, or inadequate reporting; or (c) the neoplasms produced often occur spontaneously and, in the past, have been difficult to classify as malignant by histological criteria alone (e.g., lung adenomas and adenocarcinomas and liver tumours in certain strains of mice).

(3) *Inadequate evidence* is available when, because of major qualitative or quantitative limitations, the studies cannot be interpreted as showing either the presence or absence of a carcinogenic effect.

(4) *No evidence* applies when several adequate studies are available which show that, within the limits of the tests used, the chemical or complex mixture is not carcinogenic.

It should be noted that the categories *sufficient evidence* and *limited evidence* refer only to the strength of the experimental evidence that these chemicals or complex mixtures are carcinogenic and not to the extent of their carcinogenic activity nor to the mechanism involved. The classification of any chemical may change as new information becomes available.

(ii) *Evidence for activity in short-term tests*[1]

Many short-term tests bearing on postulated mechanisms of carcinogenesis or on the properties of known carcinogens have been developed in recent years. The induction of cancer is thought to proceed by a series of steps, some of which have been distinguished experimentally (16-20). The first step - initiation - is thought to involve damage to DNA, resulting in heritable alterations in or rearrangements of genetic information. Most short-term tests in common use today are designed to evaluate the genetic activity of a substance. Data from these assays are useful for identifying potential carcinogenic hazards, in identifying active metabolites of known carcinogens in human or animal body fluids, and in helping to elucidate mechanisms of carcinogenesis. Short-term tests to detect agents with tumour-promoting activity are, at this time, insufficiently developed.

Because of the large number of short-term tests, it is difficult to establish rigid criteria for adequacy that would be applicable to all studies. General considerations relevant to all tests, however, include (a) that the test system be valid with respect to known animal carcinogens and noncarcinogens; (b) that the experimental parameters under which the chemical (or complex mixture) is tested include a sufficiently wide dose range and duration of exposure to the agent and an appropriate metabolic system; (c) that appropriate controls be used; and (d) that the purity of the compound or, in the case of complex mixtures, that the source and representativity of the sample being tested be specified. Confidence in positive results is increased if a dose-response relationship is demonstrated and if this effect has been reported in two or more independent studies.

Most established short-term tests employ as end-points well-defined genetic markers in prokaryotes and lower eukaryotes and in mammalian cell lines. The tests can be grouped according to the end-point detected:

Tests of *DNA damage*. These include tests for covalent binding to DNA, induction of DNA breakage or repair, induction of prophage in bacteria and differential survival of DNA repair-proficient/-deficient strains of bacteria.

Tests of *mutation* (measurement of heritable alterations in phenotype and/or genotype). These include tests for detection of the loss or alteration of a gene product, and change of function through forward or reverse mutation, recombination and gene conversion; they may involve the nuclear genome, the mitochondrial genome and resident viral or plasmid genomes.

Tests of *chromosomal effects*. These include tests for detection of changes in chromosome number (aneuploidy), structural chromosomal aberrations, sister chromatid exchanges, micronuclei and dominant-lethal events. This classification does not imply that some chromosomal effects are not mutational events.

Tests for *cell transformation*, which monitor the production of preneoplastic or neoplastic cells in culture, are also of importance because they attempt to simulate essential steps in

[1]Based on the recommendations of a working group which met in 1983 (5)

cellular carcinogenesis. These assays are not grouped with those listed above since the mechanisms by which chemicals induce cell transformation may not necessarily be the result of genetic change.

The selection of specific tests and end-points for consideration remains flexible and should reflect the most advanced state of knowledge in this field.

The data from short-term tests are summarized by the Working Group and the test results tabulated according to the end-points detected and the biological complexities of the test systems. The format of the table used is shown below. In these tables, a '+' indicates that the compound was judged by the Working Group to be significantly positive in one or more assays for the specific end-point and level of biological complexity; '−' indicates that it was judged to be negative in one or more assays; and '?' indicates that there were contradictory results from different laboratories or in different biological systems, or that the result was judged to be equivocal. These judgements reflect the assessment by the Working Group of the quality of the data (including such factors as the purity of the test compound, problems of metabolic activation and appropriateness of the test system) and the relative significance of the component tests.

Overall assessment of data from short-term tests

	Genetic activity			Cell transformation
	DNA damage	Mutation	Chromosomal effects	
Prokaryotes				
Fungi/ Green plants				
Insects				
Mammalian cells (*in vitro*)				
Mammals (*in vivo*)				
Humans (*in vivo*)				

An overall assessment of the evidence for *genetic activity* is then made on the basis of the entries in the table, and the evidence is judged to fall into one of four categories, defined as follows:

(i) *Sufficient evidence* is provided by at least three positive entries, one of which must involve mammalian cells *in vitro* or *in vivo* and which must include at least two of three end-points - DNA damage, mutation and chromosomal effects.

(ii) *Limited evidence* is provided by at least two positive entries.

(iii) *Inadequate evidence* is available when there is only one positive entry or when there are too few data to permit an evaluation of an absence of genetic activity or when there are unexplained, inconsistent findings in different test systems.

(iv) *No evidence* applies when there are only negative entries; these must include entries for at least two end-points and two levels of biological complexity, one of which must involve mammalian cells *in vitro* or *in vivo*.

It is emphasized that the above definitions are operational, and that the assignment of a chemical or complex mixture into one of these categories is thus arbitrary.

In general, emphasis is placed on positive results; however, in view of the limitations of current knowledge about mechanisms of carcinogenesis, certain cautions should be respected: (i) At present, short-term tests should not be used by themselves to conclude whether or not an agent is carcinogenic, nor can they predict reliably the relative potencies of compounds as carcinogens in intact animals. (ii) Since the currently available tests do not detect all classes of agents that are active in the carcinogenic process (e.g., hormones), one must be cautious in utilizing these tests as the sole criterion for setting priorities in carcinogenesis research and in selecting compounds for animal bioassays. (iii) Negative results from short-term tests cannot be considered as evidence to rule out carcinogenicity, nor does lack of demonstrable genetic activity attribute an epigenetic or any other property to a substance (5).

(b) Evaluation of Carcinogenicity in Humans

Evidence of carcinogenicity can be derived from case reports, descriptive epidemiological studies and analytical epidemiological studies.

An analytical study that shows a positive association between an exposure and a cancer may be interpreted as implying causality to a greater or lesser extent, on the basis of the following criteria: (a) There is no identifiable positive bias. (By 'positive bias' is meant the operation of factors in study design or execution that lead erroneously to a more strongly positive association between an exposure and disease than in fact exists. Examples of positive bias include, in case-control studies, better documentation of the exposure for cases than for controls, and, in cohort studies, the use of better means of detecting cancer in exposed individuals than in individuals not exposed.) (b) The possibility of positive confounding has been considered. (By 'positive confounding' is meant a situation in which the relationship between an exposure and a disease is rendered more strongly positive than it truly is as a result of an association between that exposure and another exposure which either causes or prevents the disease. An example of positive confounding is the association between coffee consumption and lung cancer, which results from their joint association with cigarette smoking.) (c) The association is unlikely to be due to chance alone. (d) The association is strong. (e) There is a dose-response relationship.

In some instances, a single epidemiological study may be strongly indicative of a cause-effect relationship; however, the most convincing evidence of causality comes when several independent studies done under different circumstances result in 'positive' findings.

Analytical epidemiological studies that show no association between an exposure and a cancer ('negative' studies) should be interpreted according to criteria analogous to those listed above: (a) there is no identifiable negative bias; (b) the possibility of negative confounding has been considered; and (c) the possible effects of misclassification of exposure or outcome have been weighed. In addition, it must be recognized that the probability that a given study can detect a certain effect is limited by its size. This can be perceived from the confidence limits around the estimate of association or relative risk. In a study regarded as 'negative', the upper confidence limit may indicate a relative risk substantially greater than unity; in that case, the study excludes only relative risks that are above the upper limit. This usually means that a 'negative' study must be large to be convincing. Confidence in a 'negative' result is increased when several independent studies carried out under different circumstances are in agreement. Finally, a 'negative' study may be considered to be relevant only to dose levels within or below the range of those observed in the study and is pertinent only if sufficient time has elapsed since first human exposure to the agent. Experience with human cancers of known etiology suggests that the period from first exposure to a chemical carcinogen to development of clinically observed cancer is usually measured in decades and may be in excess of 30 years.

The evidence for carcinogenicity from studies in humans is assessed by the Working Group and judged to fall into one of four groups, defined as follows:

1. *Sufficient evidence* of carcinogenicity indicates that there is a causal relationship between the exposure and human cancer.
2. *Limited evidence* of carcinogenicity indicates that a causal interpretation is credible, but that alternative explanations, such as chance, bias or confounding, could not adequately be excluded.
3. *Inadequate evidence*, which applies to both positive and negative evidence, indicates that one of two conditions prevailed: (a) there are few pertinent data; or (b) the available studies, while showing evidence of association, do not exclude chance, bias or confounding.
4. *No evidence* applies when several adequate studies are available which do not show evidence of carcinogenicity.

(c) Relevance of Experimental Data to the Evaluation of Carcinogenic Risk to Humans

Information compiled from the first 29 volumes of the *IARC Monographs* (4,21,22) shows that, of the chemicals or groups of chemicals now generally accepted to cause or probably to cause cancer in humans, all (with the possible exception of arsenic) of those that have been tested appropriately produce cancer in at least one animal species. For several of the chemicals (e.g., aflatoxins, 4-aminobiphenyl, diethylstilboestrol, melphalan, mustard gas and vinyl chloride), evidence of carcinogenicity in experimental animals preceded evidence obtained from epidemiological studies or case reports.

For many of the chemicals (or complex mixtures) evaluated in the *IARC Monographs* for which there is *sufficient evidence* of carcinogenicity in animals, data relating to carcinogenicity for humans are either insufficient or nonexistent. **In the absence of adequate data on humans, it is reasonable, for practical purposes, to regard chemicals for which there is sufficient evidence of carcinogenicity in animals as if they presented a carcinogenic risk to humans.** The use of the expressions 'for practical purposes' and 'as if they presented a carcinogenic risk' indicates that, at the present time, a correlation between carcinogenicity in animals and possible human risk cannot be made on a purely scientific basis, but only pragmatically. Such a pragmatical correlation may be useful to regulatory agencies in making decisions related to the primary prevention of cancer.

In the present state of knowledge, it would be difficult to define a predictable relationship between the dose (mg/kg bw per day) of a particular chemical required to produce cancer in test animals and the dose that would produce a similar incidence of cancer in humans. Some data, however, suggest that such a relationship may exist (23,24), at least for certain classes of carcinogenic chemicals, although no acceptable method is currently available for quantifying the possible errors that may be involved in such an extrapolation procedure.

8. EXPLANATORY NOTES ON THE CONTENTS OF MONOGRAPHS ON OCCUPATIONAL EXPOSURES TO COMPLEX MIXTURES

(a) Historical Perspectives (Section 1)

The origins and development of the industry are described. The number of plants, the number of workers employed in the industry and production figures are given when available. Since cancer is a delayed toxic effect, changes over time in processes or in the chemicals used which may have been significant with respect to the development of cancer are indicated.

(b) Description of the Industry (Section 2)

Processes used within the industry that are considered to be representative of those prevalent in the majority of plants are described; important variations are also outlined. It should not be inferred, however, that these are the only processes used. Processes during which exposures to solvents, fumes and dusts may occur are noted in particular.

(c) Exposures in the Workplace (Section 3)

Measurements of exposures in the workplace, as determined by area or personal sampling, are reported. When few published data are available to the Working Group, unpublished reports that are deemed to be appropriate may be included. Cross references are given to previous monographs on particular chemicals identified in the industry.

(d) Biological Data Relevant to the Evaluation of Carcinogenic Risk to Humans (Section 4)

The studies of carcinogenicity and mutagenicity summarized in section 4 include only those performed on samples of complex mixtures taken from the industry. In general, the data recorded in section 4 are summarized as given by the author; however, comments made by the Working Group on certain shortcomings of reporting, of statistical analysis or of experimental design are given in square brackets.

(i) *Carcinogenicity studies in animals*

The monographs are not intended to cover all reported studies. Some studies are purposely omitted (a) because they are inadequate, as judged from previously described criteria (25-28) (e.g., too short a duration, too few animals, poor survival); (b) because they only confirm findings that have already been fully described; or (c) because they are judged irrelevant for the purpose of the evaluation. In certain cases, however, such studies are mentioned briefly, particularly when the information is considered to be a useful supplement to other reports or when it is the only data available. Their inclusion does not, however, imply acceptance of the adequacy of their experimental design or of the analysis and interpretation of their results.

Mention is made of all routes of administration by which the test material has been adequately tested and of all species in which relevant tests have been done (28). In most cases, animal strains are given. Quantitative data are given to indicate the order of magnitude of the effective carcinogenic doses. In general, the doses and schedules are indicated as they appear in the original; sometimes units have been converted for easier comparison.

(ii) *Other relevant biological data*

Data on toxicity, on effects on reproduction, on teratogenicity and embryo- and fetotoxicity and on placental transfer, from studies in experimental animals and from observations in humans, are included when considered relevant.

Data from short-term tests are also included. In addition to the tests for genetic activity and cell transformation described previously (see pages 16-18), data from studies of related effects, but for which the relevance to the carcinogenic process is less well established, may also be mentioned.

The criteria used for considering short-term tests and for evaluating their results have been described (see pages 16-18). In general, the authors' results are given as reported. An assessment of the data by the Working Group which differs from that of the authors, and comments concerning aspects of the study that might affect its interpretation are given in square brackets. Reports of studies in which few or no experimental details are given, or in which the data on which a reported positive or negative result is based are not available for examination, are cited, but are identified as 'abstract' or 'details not given' and are not considered in the summary tables or in making the overall assessment of genetic activity.

For several recent reviews on short-term tests, see IARC (28), Montesano *et al.* (29), de Serres and Ashby (30), Sugimura *et al.* (31), Bartsch *et al.* (32) and Hollstein *et al.* (33).

(iii) *Case reports and epidemiological studies of carcinogenicity to humans*

Observations in humans are summarized in this section. These include case reports, descriptive epidemiological studies (which correlate cancer incidence in space or time to an exposure) and analytical epidemiological studies of the case-control or cohort type. In principle, a comprehensive coverage is made of observations in humans; however, reports are excluded when judged to be clearly not pertinent. This applies in particular to case reports, in which either the clinico-pathological description of the tumours or the exposure history, or both, are poorly described; and to published routine statistics, for example, of cancer mortality by occupational category, when the categories are so broadly defined as to contribute virtually no specific information on the possible relation between cancer occurrence and a given exposure. Results of studies are assessed on the basis of the data and analyses that are presented in the published papers. Some additional analyses of the published data may be performed by the Working Group to gain better insight into the relation between cancer occurrence and the exposure under consideration. The Working Group may use these analyses in its assessment of the evidence or may actually include them in the text to summarize a study; in such cases, the results of the supplementary analyses are given in square brackets. Any comments by the Working Group are also reported in square brackets; however, these are kept to a minimum, being restricted to those instances in which it is felt that an important aspect of a study, directly impinging on its interpretation, should be brought to the attention of the reader.

(*d*) *Summary of Data Reported and Evaluation (Section 5)*

Section 5 summarizes the relevant data from animals and humans and gives the critical views of the Working Group on those data.

(i) *Exposures*

A very brief description of the industry is given, and exposures that are considered pertinent to an assessment of human carcinogenicity are cited.

(ii) *Experimental data*

Data relevant to the evaluation of the carcinogenicity of the test material in animals are summarized in this section. The animal species mentioned are those in which the carcinogenicity of the substance was clearly demonstrated. Tumour sites are also indicated. If the substance has produced tumours after prenatal exposure or in single-dose experiments, this is indicated. Dose-response data are given when available.

Significant findings on effects on reproduction and prenatal toxicity, and results from short-term tests for genetic activity and cell transformation assays are summarized, and the latter are presented in tables. An overall assessment of the degree of evidence for genetic activity in short-term tests is made.

(iii) *Human data*

Case reports and epidemiological studies that are considered to be pertinent to an assessment of human carcinogenicity are described. Other biological data that are considered to be relevant are also mentioned.

(iv) *Evaluation*

This section comprises the overall evaluation by the Working Group of the carcinogenic risk of the occupational exposure to humans. All of the data in the monograph, and particularly the summarized information on human and experimental carcinogenicity, are considered in order to make the evaluation. This section should be read in conjunction with pages 15 and 19 of this Preamble for definitions of degrees of evidence.

References

1. IARC (1979) Criteria to select chemicals for *IARC Monographs*. *IARC intern. tech. Rep. No. 79/003*

2. IARC (1977) IARC Monograph Programme on the Evaluation of the Carcinogenic Risk of Chemicals to Humans. Preamble. *IARC intern. tech. Rep. No. 77/002*

3. IARC (1978) Chemicals with *sufficient evidence* of carcinogenicity in experimental animals - *IARC Monographs* volumes 1-17. *IARC intern. tech. Rep. No. 78/003*

4. IARC (1982) *IARC Monographs on the Evaluation of the Carcinogenic Risk of Chemicals to Humans*, Supplement 4, *Chemicals, Industrial Processes and Industries Associated with Cancer in Humans* (IARC Monographs Volumes 1 to 29)

5. IARC (1983) Approaches to classifying chemical carcinogens according to mechanism of action. *IARC intern. tech. Rep. No. 83/001*

6. IARC (1972-1983) *IARC Monographs on the Evaluation of the Carcinogenic Risk of Chemicals to Humans*, Volumes 1-34, Lyon, France

   Volume 1 (1972) Some Inorganic Substances, Chlorinated Hydrocarbons, Aromatic Amines, *N*-Nitroso Compounds and Natural Products (19 monographs), 184 pages

   Volume 2 (1973) Some Inorganic and Organometallic Compounds (7 monographs), 181 pages

   Volume 3 (1973) Certain Polycyclic Aromatic Hydrocarbons and Heterocyclic Compounds (17 monographs), 271 pages

   Volume 4 (1974) Some Aromatic Amines, Hydrazine and Related Substances, *N*-Nitroso Compounds and Miscellaneous Alkylating Agents (28 monographs), 286 pages

   Volume 5 (1974) Some Organochlorine Pesticides (12 monographs), 241 pages

   Volume 6 (1974) Sex Hormones (15 monographs), 243 pages

   Volume 7 (1974) Some Anti-thyroid and Related Substances, Nitrofurans and Industrial Chemicals (23 monographs), 326 pages

   Volume 8 (1975) Some Aromatic Azo Compounds (32 monographs), 357 pages

   Volume 9 (1975) Some Aziridines, *N*-, *S*- and *O*-Mustards and Selenium (24 monographs), 268 pages

   Volume 10 (1976) Some Naturally Occurring Substances (22 monographs), 353 pages

   Volume 11 (1976) Cadmium, Nickel, Some Epoxides, Miscellaneous Industrial Chemicals and General Considerations on Volatile Anaesthetics (24 monographs), 306 pages

   Volume 12 (1976) Some Carbamates, Thiocarbamates and Carbazides (24 monographs), 282 pages

Volume 33 (1984) Polynuclear Aromatic Compounds, Part 2, Carbon Blacks, Mineral Oils and Some Nitroarenes (8 monographs), 245 pages

Volume 34 (1984) Polynuclear Aromatic Compounds, Part 3, Industrial Exposures in Aluminium Production, Coal Gasification, Coke Production, and Iron and Steel Founding (4 monographs), 219 pages

Supplement No. 1 (1979) Chemicals and Industrial Processes Associated with Cancer in Humans (IARC Monographs, Volumes 1 to 20), 71 pages

Supplement No. 2 (1980) Long-term and Short-term Screening Assays for Carcinogens: A Critical Appraisal, 426 pages

Supplement No. 3 (1982) Cross Index of Synonyms and Trade Names in Volumes 1 to 26, 199 pages

Supplement No. 4 (1982) Chemicals, Industrial Processes and Industries Associated with Cancer in Humans (IARC Monographs, Volumes 1 to 29), 292 pages

7. IARC (1973-1983) *Information Bulletin on the Survey of Chemicals Being Tested for Carcinogenicity*, Numbers 1-10, Lyon, France

   Number 1 (1973) 52 pages
   Number 2 (1973) 77 pages
   Number 3 (1974) 67 pages
   Number 4 (1974) 97 pages
   Number 5 (1975) 88 pages
   Number 6 (1976) 360 pages
   Number 7 (1978) 460 pages
   Number 8 (1979) 604 pages
   Number 9 (1981) 294 pages
   Number 10 (1983) 326 pages

8. PHS 149 (1951-1983) Public Health Service Publication No. 149, *Survey of Compounds which have been Tested for Carcinogenic Activity*, Washington DC, US Government Printing Office

   1951 Hartwell, J.L., 2nd ed., Literature up to 1947 on 1329 compounds, 583 pages
   1957 Shubik, P. & Hartwell, J.L., Supplement 1, Literature for the years 1948-1953 on 981 compounds, 388 pages
   1969 Shubik, P. & Hartwell, J.L., edited by Peters, J.A., Supplement 2, Literature for the years 1954-1960 on 1048 compounds, 655 pages
   1971 National Cancer Institute, Literature for the years 1968-1969 on 882 compounds, 653 pages
   1973 National Cancer Institute, Literature for the years 1961-1967 on 1632 compounds, 2343 pages
   1974 National Cancer Institute, Literature for the years 1970-1971 on 750 compounds, 1667 pages
   1976 National Cancer Institute, Literature for the years 1972-1973 on 966 compounds, 1638 pages
   1980 National Cancer Institute, Literature for the year 1978 on 664 compounds, 1331 pages
   1983 National Cancer Institute, Literature for years 1974-1975 on 575 compounds, 1043 pages

9. Pike, M.C. & Roe, F.J.C. (1963) An actuarial method of analysis of an experiment in two-stage carcinogenesis. *Br. J. Cancer*, *17*, 605-610

10. Miller, E.C. (1978) Some current perspectives on chemical carcinogenesis in humans and experimental animals: Presidential address. *Cancer Res.*, *38*, 1479-1496

11. Miller, E.C. & Miller, J.A. (1981) Searches for ultimate chemical carcinogens and their reactions with cellular macromolecules. *Cancer*, *47*, 2327-2345

12. Peto, R. (1974) Guidelines on the analysis of tumour rates and death rates in experimental animals. *Br. J. Cancer*, *29*, 101-105

13. Peto, R. (1975) Letter to the editor. *Br. J. Cancer*, *31*, 697-699

14. Hoel, D.G. & Walburg, H.E., Jr (1972) Statistical analysis of survival experiments. *J. natl Cancer Inst.*, *49*, 361-372

15. Peto, R., Pike, M.C., Day, N.E., Gray, R.G., Lee, P.N., Parish, S., Peto, J., Richards, S. & Wahrendorf, J. (1980) *Guidelines for simple sensitive significance tests for carcinogenic effects in long-term animal experiments.* In: *IARC Monographs on the Evaluation of the Carcinogenic Risk of Chemicals to Humans, Supplement 2, Long-term and Short-term Screening Assays for Carcinogens: A Critical Appraisal*, Lyon, pp. 311-426

16. Berenblum, I. (1975) *Sequential aspects of chemical carcinogenesis: Skin.* In: Becker, F.F., ed., *Cancer. A Comprehensive Treatise*, Vol. 1, New York, Plenum Press, pp. 323-344

17. Foulds, L. (1969) *Neoplastic Development*, Vol. 2, London, Academic Press

18. Farber, E. & Cameron, R. (1980) The sequential analysis of cancer development. *Adv. Cancer Res.*, *31*, 125-126

19. Weinstein, I.B. (1981) The scientific basis for carcinogen detection and primary cancer prevention. *Cancer*, *47*, 1133-1141

20. Slaga, T.J., Sivak, A. & Boutwell, R.K., eds (1978) *Mechanisms of Tumor Promotion and Cocarcinogenesis*, Vol. 2, New York, Raven Press

21. IARC Working Group (1980) An evaluation of chemicals and industrial processes associated with cancer in humans based on human and animal data: *IARC Monographs* Volumes 1 to 20. *Cancer Res.*, *40*, 1-12

22. IARC (1979) *IARC Monographs on the Evaluation of the Carcinogenic Risk of Chemicals to Humans*, Supplement 1, *Chemicals and Industrial Processes Associated with Cancer in Humans*, Lyon

23. Rall, D.P. (1977) *Species differences in carcinogenesis testing.* In: Hiatt, H.H., Watson, J.D. & Winsten, J.A., eds, *Origins of Human Cancer*, Book C, Cold Spring Harbor, NY, Cold Spring Harbor Laboratory, pp. 1383-1390

24. National Academy of Sciences (NAS) (1975) *Contemporary Pest Control Practices and Prospects: The Report of the Executive Committee*, Washington DC

25. WHO (1958) Second Report of the Joint FAO/WHO Expert Committee on Food Additives. Procedures for the testing of intentional food additives to establish their safety and use. *WHO tech. Rep. Ser., No. 144*

26. WHO (1967) Scientific Group. Procedures for investigating intentional and unintentional food additives. *WHO tech. Rep. Ser., No. 348*

27. Sontag, J.M., Page, N.P. & Saffiotti, U. (1976) Guidelines for carcinogen bioassay in small rodents. *Natl Cancer Inst. Carcinog. tech. Rep. Ser., No.1*

28. IARC (1980) *IARC Monographs on the Evaluation of the Carcinogenic Risk of Chemicals to Humans*, Supplement 2, *Long-term and Short-term Screening Assays for Carcinogens: A Critical Appraisal*, Lyon

29. Montesano, R., Bartsch, H. & Tomatis, L., eds (1980) *Molecular and Cellular Aspects of Carcinogen Screening Tests* (*IARC Scientific Publications No. 27*), Lyon

30. de Serres, F.J. & Ashby, J., eds (1981) *Evaluation of Short-Term Tests for Carcinogens. Report of the International Collaborative Program*, Amsterdam, Elsevier/North-Holland Biomedical Press

31. Sugimura, T., Sato, S., Nagao, M., Yahagi, T., Matsushima, T., Seino, Y., Takeuchi, M. & Kawachi, T. (1976) *Overlapping of carcinogens and mutagens*. In: Magee, P.N., Takayama, S., Sugimura, T. & Matsushima, T., eds, *Fundamentals in Cancer Prevention*, Tokyo/Baltimore, University of Tokyo/University Park Press, pp. 191-215

32. Bartsch, H., Tomatis, L. & Malaveille, C. (1982) *Qualitative and quantitative comparison between mutagenic and carcinogenic activities of chemicals*. In: Heddle, J.A., ed., *Mutagenicity: New Horizons in Genetic Toxicology*, New York, Academic Press, pp. 35-72

33. Hollstein, M., McCann, J., Angelosanto, F.A. & Nichols, W.W. (1979) Short-term tests for carcinogens and mutagens. *Mutat. Res.*, *65*, 133-226

GENERAL REMARKS ON THE INDUSTRIAL EXPOSURES CONSIDERED

This thirty-fourth volume of *IARC Monographs* is the third on industrial exposures: monographs were prepared previously on occupational exposures as they occur in the wood, leather and associated manufacturing industries (IARC, 1981) and the rubber manufacturing industry (IARC, 1982).

1. *Choice of industries and scope of coverage*

The present volume considers some industries in which exposure to polynuclear aromatic compounds (PACs) may occur - aluminium production, coal gasification, coke production, and iron and steel founding, and is the third in a series of four volumes dealing with PACs. Volume 32 (IARC, 1983) of the *Monographs* considered chemical, environmental and experimental data on 48 individual PACs which occur in complex mixtures resulting from the combustion and pyrolysis of fossil fuels or products derived from them. In Volume 33 (IARC, 1984), experimental and epidemiological data on carbon blacks, mineral oils and some nitroarenes were considered.

The four industries considered in the present volume were chosen on the basis of known exposures to PACs from coal- and petroleum-derived substances used in these industries and the availability of epidemiological studies suggesting a potential cancer hazard. An additional consideration was the fact that, worldwide, these industries employ more than two million people. Environmental exposures that occur outside the workplace are not considered in these monographs.

The monograph on aluminium production covers only the electrolytic reduction of alumina to aluminium and the casting of aluminium into ingots and thus does not include bauxite mining, alumina extraction or founding and aluminium processing works.

In the epidemiological studies reported in the early literature on coal carbonization, data on health effects from the process of coal gasification could not be distinguished from those from coke production; consequently, these early studies are evaluated separately. Most of the later epidemiological data, however, distinguish between workers employed in the production of town gas and workers employed at coke-oven plants in steel works where coke is the primary end product and where the gas, produced as a by-product, is used within the plant or locally. These studies are dealt with in the monographs on coal gasification and coke production, respectively.

Gasification entails either the destructive distillation of coal or the treatment of coal in a reducing atmosphere (produced by partial oxidation of coal with oxygen, air and steam) to yield a combustible gaseous product. Data on working conditions and health effects in the newer processes, such as Lurgi gasifiers, are fragmentary. Coal liquefaction processes, to produce liquid hydrocarbons by hydrogenation, are not considered.

The monograph on coke production is confined principally to slot-oven coke batteries and, therefore, is not representative of all processes, nor does it include data on all the possible products, by-products, and emissions from coal conversion.

The iron and steel founding industry is defined here as starting with the melting of alloys and ending with the fettling of castings. Shaped castings manufactured in the foundries are distinguished from the ingots and other cast forms that are produced in iron and steel works outside the founding industry. The founding industry as defined comprises the following basic operations: patternmaking, moulding, core-making, melting and pouring, shake-out and fettling.

A glossary of terms used in the four industries considered is provided at the end of this volume.

Carcinogenicity studies in animals and studies on mutagenicity and related effects reported in these monographs include only those on samples of complex mixtures taken from the industry. Studies pertaining to crude coal-tars, as they might occur at the workplaces in question, have also been included. The tars were obtained from either gas retort houses or coke-oven plants, and the studies are reported in the relevant monograph.

Epidemiological data on tar distillery workers have not been summarized, since such data, together with all epidemiological and experimental data on pharmaceutical coal-tars and coal-tar distillation products such as creosotes and coal-tar pitches will be considered by a Working Group convening in February 1984. That meeting will be the fourth in the series related to PACs and will also cover bitumens (asphalts), shale oils and chimney soots from domestic and institutional sources.

2. *Evolution of industries over time*

About 40 countries produce aluminium, all using basically the same electrolytic reduction method but various processes. Since the development of electrolytic reduction, the production of aluminium has been increasing steadily, and no change in this trend is expected in the near future, although the technology has continued to evolve.

Town gas was produced widely by the destructive distillation of coal from the latter part of the nineteenth century until the 1960s, and it is still manufactured in that way in a number of countries. Since the 1920s, a number of other coal gasification technologies have been developed. Currently, there are several types of modern coal-gasification plants and experimental gasifiers in operation throughout the world.

The production of coke and coke-oven gas was widespread throughout the nineteenth and twentieth centuries. At the beginning of this century, most coke was produced in 'beehive' ovens. As the demand for coke increased and a commercial use for coke-oven gas was developed, the slot-oven, in which volatiles are collected, became used more widely. With the decline in the use of coke-oven gas for non-industrial purposes and its increased production from slot-ovens, the numbers of workers and coke batteries have decreased throughout the world. However, the manufacture of coke is a necessary process in the metallurgical industry, and its production is expected to remain constant in the future.

The iron and steel founding industry is spread throughout industrialized countries. The output of this industry remained constant for many years. Recent changes in the industry would suggest that, although the number of workers engaged in iron and steel founding may not increase in the future, owing to higher productivity, continuing technological developments may change the exposures of the workers, with the introduction of new chemicals, particularly in the moulding department. The founding industry, and the working conditions therein, have changed technologically, with wide variation within and between countries, and methods ranging from manual to fully automated processes continue to be in operation.

3. *Problems in determining the relation between exposure and disease*

The search for carcinogenic agents that could account for an excess of a common tumour such as lung cancer in these complex industries is a much more difficult task than that in situations in which an excess of a rare type of cancer has been noted, e.g., liver haemangiosarcomas and vinyl chloride. In addition to the large number of agents to which workers may be exposed in the industries concerned, there are other variables that make it difficult to assess the exposure. The changes over time in the chemicals used and the processes employed may change the exposure spectrum, making it even more difficult to interpret exposure-effect relationships. In addition, there is a lack of environmental monitoring data, particularly from earlier years, and epidemiological studies rarely describe the process employed. Epidemiological studies have had to rely primarily on occupational classifications which do not take into account exposures. These titles are usually poorly defined in relation to exposure in industrial settings, and individual workers may change their jobs and even industries throughout their working life.

In view of the complexity of the airborne contaminants in the industries being considered, it is not clear which of the components of the tars are the most important to measure in order to protect the health of employees. Some environmental criteria are based on analysis of benzene-soluble fractions (pitch volatiles), which contain PACs and azaarenes. However, because the method is non-specific and any material soluble in benzene or cyclohexane is measured (Walker, 1977; McEachen *et al.*, 1978), the approach has recently come under scrutiny.

In some studies, a single PAC (usually benzo[*a*]pyrene) has been measured as an index of PAC exposure. As analytical procedures have improved, the additive total of individual PACs has also been used as an index of such exposure. This approach has the advantages of specifically including PACs in the analysis (rather than benzene- or cyclohexane-soluble substances in general) and of measuring more than one PAC. However, it may be limited by the analytical methodology available, by costs and by the weighting it gives to individual PACs. As more is learned about PAC profiles for different operations and about the relative toxicities of individual PACs, more selective analytical procedures may evolve. However, at present, the relationship of any of these approaches for assessing PAC exposures to actual cancer risk for exposed workers is uncertain other than as a relative indicator of high *versus* low PAC exposure.

Occupational exposure limits for PACs are shown in Table 1.

Table 1. Occupational exposure limits for coal-tar pitch volatiles (CTPV, as the benzene-soluble fraction of total particulate matter) and for benzo[*a*]pyrene (B[*a*]P) in different countries

Country	Substance	Exposure limit (μg/m^3)
Australia[a]	CTPV	200
Belgium[a]	CTPV	200
Finland[b]	B[*a*]P	10
Italy[a]	CTPV	200
The Netherlands[a]	CTPV	200
Sweden[c]	B[*a*]P	5
Switzerland[a]	B[*a*]P	200
USA		
NIOSH[d]	CTPV	100
OSHA[e]	CTPV	150
ACGIH[f]	CTPV	200
USSR[a]	B[*a*]P	0.15
Yugoslavia[a]	CTPV	200

[a]International Labour Office (1980)
[b]National Finnish Board of Occupational Safety and Health (1981)
[c]National Swedish Board of Occupational Safety and Health (1981)
[d]National Institute for Occupational Safety and Health (1977)
[e]US Occupational Safety and Health Administration (1976)
[f]American Conference of Governmental Industrial Hygienists (1983)

4. *Problems in evaluating experimental results*

The variability of the chemical composition of the emissions of tarry products in each of the four industries considered presents a major difficulty in making a general evaluation of the studies of experimental carcinogenicity and mutagenicity in relation to the whole industry. The Working Group stressed, therefore, that the results of such experiments are valid only for the samples tested; it should not be implied that the results from a single sample would be valid for all samples taken from other parts of the industry or from the same location but at a later date. However, experimental carcinogenicity studies are important and have been instrumental in differentiating the relative carcinogenicity of various fractions of tars originating from coal (Berenblum & Schoental, 1947).

In addition to environmental monitoring, biological monitoring has been used to measure exposure, e.g., mutagenic activity as well as levels of individual PACs or their metabolites have been determined in workers' urine. However, there continue to be difficulties in interpreting such results.

5. *Problems in interpreting epidemiological studies*

Specific limitations to the epidemiological data used to assess industries are common to all of the industries reviewed in this volume and to occupational studies in general.

(*a*) *Types of epidemiological studies*

As is true in any assessment of carcinogenesis, the design of the epidemiological studies used may vary, but, in assessing occupational effects, this variation may be more of a problem because there are so many other variables associated with industrial exposures. In the descriptive studies, the distribution of occupations on the death certificates of cancer patients is compared to the occupational data from a census, and these two sources of data have been shown to be inconsistent. In case-control studies, relatives usually cannot precisely define the job of a deceased patient.

The proportional mortality ratio (PMR; the proportion of all deaths due to a specific cause) cannot be equated with an estimate of risk. Biases may occur in such analyses because the total number of deaths in a population is often not included; and the measure itself is misleading, since an increase or decrease in the proportion of one cause of death has a direct influence, in the opposite direction, on the proportion of other causes of death.

Even the cohort studies are often difficult to evaluate when they are limited in number, contain overlapping populations or show differences in risk estimates depending on the population of industrial workers selected and the comparison group.

Most epidemiological studies referred to in this volume focus only on mortality; since there were no morbidity studies, cancers, such as skin cancers, which might be a problem in industries with coal-tar exposures, could not be assessed epidemiologically. Industrial groups who receive good medical care may actually have prolonged survival from cancers, such as those of the bladder, making it possible to underestimate the cancer risk if only mortality data are considered.

(*b*) *Assessment of exposure*

Jobs are the only indicator of exposure available in most epidemiological studies. However, workers change jobs within the industry and it is difficult to determine how to 'sum' an individual exposure. There has been variability between studies and even different methods used within a study with different results. Exposure may be counted as 'ever' worked in a suspect area, worked in the area for a set duration of time, e.g., five years or more, or worked in the area at a calendar point in time. In addition, the division of study groups into individual job classifications may result in numbers of cause-specific deaths that are too small to assess adequately.

(*c*) *Confounding factors*

In many of these studies, lung cancer has been the observed risk and yet no information on smoking is available. If the risk is high, the lack of data on smoking may be irrelevant. However, when the risk is borderline or changes from study to study, information on smoking is needed. Ethnicity and socioeconomic factors may also be important variables, especially in workers with digestive-tract cancers. Attempts to correct for these factors through internal comparisons may limit the extent of some of these problems.

6. *References*

American Conference of Governmental Industrial Hygienists (1983) *TLVs Threshold Limit Values for Chemical Substances in the Work Environment Adopted by ACGIH for 1983-84*, Cincinnati, OH, p. 42

Berenblum, I. & Schoental, R. (1947) Carcinogenic constituents of coal-tar. *Br. J. Cancer*, *1*, 157-165

IARC (1981) *IARC Monographs on the Evaluation of the Carcinogenic Risk of Chemicals to Humans*, Vol. 25, *Wood, Leather and Some Associated Industries*, Lyon

IARC (1982) *IARC Monographs on the Evaluation of the Carcinogenic Risk of Chemicals to Humans*, Vol. 28, *The Rubber Industry*, Lyon

IARC (1983) *IARC Monographs on the Evaluation of the Carcinogenic Risk of Chemicals to Humans*, Vol. 32, *Polynuclear Aromatic Compounds, Part 1, Chemical, Environmental and Experimental Data*, Lyon

IARC (1984) *IARC Monographs on the Evaluation of the Carcinogenic Risk of Chemicals to Humans*, Vol. 33, *Polynuclear Aromatic Compounds, Part 2, Carbon Blacks, Mineral Oils and Some Nitroarenes*, Lyon

International Labour Office (1980) *Occupational Exposure Limits for Airborne Toxic Substances*, 2nd (rev.) ed. (*Occupational Safety and Health Series No. 37*), Geneva, pp. 48-49, 76-77

MacEachen, W.L., Boden, H. & Larivière, C. (1978) *Composition of benzene soluble matter collected at aluminum smelter workplace*. In: *Light Metals*, Vol. 2, New York, American Institute of Mining, Metallurgical and Petroleum Engineers (AIME), pp. 509-517

National Finnish Board of Occupational Safety and Health (1981) *Airborne Contaminants in the Workplaces* (*Safety Bulletin 3*), Helsinki, p. 8

National Institute for Occupational Safety and Health (1977) *Criteria for a Recommended Standard... Occupational Exposure to Coal Tar Products* (*DHEW (NIOSH) Publ. No. 78-107*), Washington DC, US Government Printing Office, p. 22

National Swedish Board of Occupational Safety and Health (1981) *Limit Values*, Stockholm, p. 10

US Occupational Safety and Health Administration (1976) Exposure to coke oven emissions. *Fed. Regist.*, *41*, 46751

Walker, T.J. (1977) *Hygienic aspects of aluminium reduction*. In: Hughes, J.P., ed., *Health Protection in Primary Aluminium Production*, London, International Primary Aluminium Institute, pp. 23-29

THE MONOGRAPHS

ALUMINIUM PRODUCTION

The aluminium production industry as referred to in this monograph includes those processes involved in the electrolytic reduction of alumina to aluminium and the casting of aluminium into ingots. The mining of bauxite, production of alumina from bauxite, alloying and fabrication of sheet metal, wire, foil and other such products are not considered.

1. Historical Perspectives

This subject has been reviewed by Pearson (1955) and the Aluminum Company of Canada (ALCAN) (1976, 1978).

Although the element aluminium was discovered by Sir Humphrey Davy in 1808, it was not until 1825 that Oersted produced a tiny quantity of the metal. The key development that led to the modern commercial production of aluminium took place in 1886 as the result of work undertaken independently by Hall in the USA and Héroult in France, who developed a process for the electrolytic reduction of alumina (Al_2O_3) to produce aluminium (Taylor, 1978). As alumina has a melting-point close to 2000°C, the problem was how to render alumina into a molten state such that an electrolytic method could be used. This was accomplished by dissolving alumina in molten cryolite (Na_3AlF_6) at a temperature below 1000°C; cryolite, which melts at a temperature below 1000°C, can dissolve approximately 8% of alumina. In the electrolysis, aluminium migrates to the cathode, and oxygen is released at the anode. This same process is used today and is referred to as the Hall-Héroult process.

Aluminium was first produced on a commercial scale in 1888 in France, Switzerland and the USA (ALCAN, 1978). By 1896, aluminium was being produced in Scotland (Taylor, 1978) and, by 1901, in Canada (ALCAN, 1976). During the First World War, world output increased from 70 800 tonnes in 1914 to 132 500 tonnes in 1919. Production was again stimulated at the start of the Second World War. Between 1939 and 1943, worldwide production of aluminium increased from 704 000 to 1 950 000 tonnes (ALCAN, 1978). Aluminium is currently produced in over 40 countries (see Table 1). In 1982, world production exceeded 13.5 million tonnes, down from over 15.5 in 1981, with the major producers being (production in million tonnes): the USA (3.3), the USSR (2.4), Canada (1.1), the Federal Republic of Germany (0.7), Norway (0.6), Australia (0.4), France (0.4), Japan (0.4), the People's Republic of China (0.4) and Spain (0.4) (American Bureau of Metal Statistics, 1983).

The uses of aluminium are dictated by its chemical and physical properties, which include low specific gravity, high tensile strength, ductility, malleability, corrosion resistance and high electrical conductivity. It is used widely, alone or in alloys, in construction (e.g., window frames, doors, building facings); transportation (e.g., automobiles, aeroplanes, trains, ships); transmission of electricity (e.g., transmission lines); machinery and equipment (e.g., cranes, scaffolding); and packaging (e.g., kitchen foils).

Table 1. Countries producing primary aluminium[a]

Country	Total 1982 production (thousand tonnes)
USA	3248
USSR	2381
Canada	1056
Federal Republic of Germany	720
Norway	632
France	387
Spain	362
Australia	359
Japan	348
People's Republic of China	347
Brazil	270
The Netherlands	249
United Kingdom	239
Italy	230
Romania	228
India	215
Venezuela	180
Yugoslavia	180
Ghana	173
Bahrain	170
New Zealand	163
Egypt	142
Argentina	136
Greece	134
Dubai	106
Republic of South Africa	99
Austria	94
United Republic of Cameroon	78
Sweden[b]	77
Iceland	75
Hungary	74
Switzerland	75
Poland	66
German Democratic Republic	60
Surinam[c]	60
Mexico	41
Czechoslovakia	37
Turkey	32
Iran	23
Republic of Korea	15
Democratic People's Republic of Korea	10
Taiwan	10

[a]From American Bureau of Metal Statistics (1983)
[b]Includes alloys
[c]Exports

2. Description of the Industry

Descriptions of aluminium production processes are given by Pearson (1955), Hughes (1977), the International Primary Aluminium Institute (1982) and Sheehy (1983)[a].

2.1 Introduction

Aluminium is the third most common element and accounts for about 7.45% of the lithosphere. It does not occur in the free state in nature but is found in combination with oxygen, silicon and fluorine in minerals making up igneous, metamorphic and sedimentary rocks. The main industrial raw material from which aluminium is extracted is bauxite, a mixture of hydrated aluminium oxides together with small amounts of oxides of iron, silicon and titanium. Other materials used in the production of aluminium are cryolite, which is mined (in Greenland only) or manufactured synthetically, aluminium fluoride, which comes from various sources, pitches (coal- or petroleum-based) and petroleum coke.

The basic steps in aluminium production are the electrolytic reduction of alumina and casting. The process is shown diagramatically in Figure 1.

Fig. 1. Schematic diagram of the aluminium production process (from ALCAN, 1978)

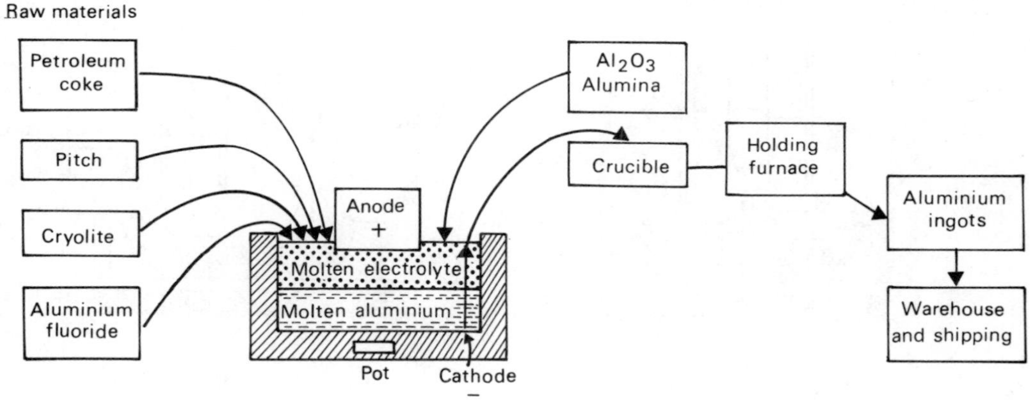

2.2 Primary aluminium production processes

(a) *Electrolytic reduction*

The process of transforming alumina into aluminium is normally referred to as the reduction process and is carried out in electrolytic cells (pots). In an aluminium reduction plant, rows of

[a]This document became available to the Secretariat subsequent to the meeting of the Working Group and hence data contained therein have not been cited in this monograph.

pots are connected in series in a potroom. A pot operating at a current of 150 000 A is common today and will yield a little over one tonne of aluminium per day. A plant might have 250 pots of this size, producing 100 000 tonnes of aluminium per year (Ravier, 1977).

There are two basic types of pot in use, prebake and Söderberg, but within each category there may be some differences in pot design. A plant may operate more than one type of pot. The basic principle is that alumina is dissolved in an electrolyte (cryolite) at a temperature of 950-970°C at a low voltage (approximately 4-6 volts), but high current is applied to the melt. The alumina is reduced to aluminium, which sinks to the bottom of the electrolytic cell and is then removed by siphoning. The oxygen from the alumina passes to the carbon anodes, where it reacts to form carbon dioxide and monoxide. The lining of the pot serves as the cathode.

(i) *Prebake anode cells*

Typical prebake pots are shown in Figures 2a and 2b.

Fabrication of anodes. For prebake pots, the anodes are fabricated in a carbon plant which is separate from the potrooms. In the green-carbon section, coke (usually calcined petroleum coke) is ground to a specific mesh size and blended with a binder of hot pitch to form a

Fig. 2. Types of electrolytic cell used in alumina reduction (from Ravier, 1977)

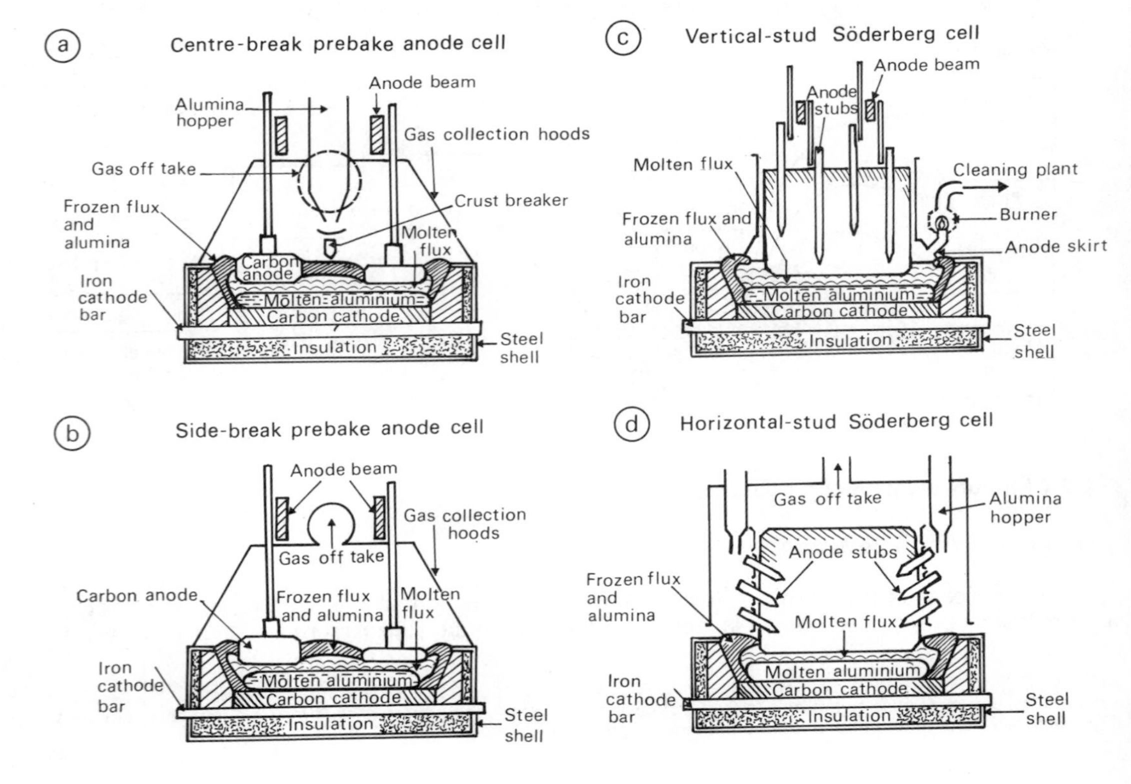

semi-solid mixture, which is then pressure moulded. At some plants this process is largely automated, while at others employees can be exposed to the dust generated in the grinding and mixing operations and to pitch volatiles. Exposure to oil mists from lubricants may occur during the anode-block pressing operation (International Primary Aluminium Institute, 1982).

The green-carbon anode is packed in ground coke and baked for several days in furnaces at approximately 1100°C. The volatiles released are either contained under negative pressure in the ovens where they are burned as fuel (International Primary Aluminium Institute, 1982), or removed and passed to scrubbers. Employees working here may be exposed to sulphur dioxide, pitch volatiles and particulates, carbon monoxide, carbon dust, dust from refractories and sometimes fluorides, if reclaimed anode butts constitute a portion of the green-carbon mixture. However, except for coke dust, occupational exposures in the carbon-baking area are minimal under normal conditions.

After baking, these carbon-anode blocks are moved to the anode rodding section, where they are fitted on rods. The rods are mated with the anode using cast iron poured into a hole in the anode. This can involve exposure of the operator to iron oxide fumes. In some plants, the anode blocks are coated with sprayed aluminium. Rods from used anodes are reconditioned in this section. This involves removal of the rod from the spent anode, straightening and cleaning. Operators may be exposed to dust and fluorides.

Operation of prebake pots. There are two designs of prebake cell. In centre-break cells (see Figure 2a), the alumina is fed into the cell either continuously or at intervals of three to six hours. A hard crust, which forms on the melt, is broken in the middle of the cell, and the hood need not be opened. In the side-break type (see Figure 2b) the alumina is added, at intervals of three to six hours, between the anodes and the wall of the cell, and the hood must be opened for crust breaking. The equipment used in some plants since the early 1950s permits the worker to perform this operation at a distance from the pot, reducing exposure to heat and fluoride fumes.

When the carbon anodes in a prebake pot are almost completely burnt out, they are replaced by a new anode assembly. Between 10 and 30 assemblies may be installed in a single pot, each lasting 20-30 days (Ravier, 1977). The anodes are replaced by a carbon setting crew. The work of the carbon setters and pot operators ('potmen') may vary from plant to plant, depending on the extent to which the process is automated, and can include the manual lowering of the anodes, adjustments of pot voltages and addition of alumina and fluoride compounds as required. Molten aluminium is siphoned from the pot under vacuum into a large crucible by a tapping crew and transported by overhead cranes to holding furnaces in the casting area.

(ii) *Söderberg anode cells*

Typical Söderberg anode cells are shown in Figure 2c and 2d. These differ from the prebake cells in that the anode is baked *in situ*.

Söderberg anode paste production. The crushing, sizing, weighing and mixing of the carbonaceous raw materials are done very much as it is for prebaked anode block production. However, the proportion and softening point of the pitch used and the temperature at which the mixers are operated is different. The warm paste discharged from the mixers may be taken directly to the Söderberg potrooms or formed into briquettes, cooled and distributed to the pots when needed.

Operation of Söderberg pots. The mixture of coke and pitch enters an anode casing above the pot, and the anode is baked by the heat produced by the passage of the electric current. In one design, studs (pins) are inserted vertically into the anode and the current enters by these studs. These pots are known as vertical-stud Söderberg pots and have only one large anode. The gases pass to the side of the anode under a skirt and are burnt off (see Figure 2c). It has been estimated that, in the best cases, 5% and, in the worst cases, 40% of the fumes, pass into the potroom. Secondary hoods can be used to improve the efficiency of fume collection to 90-95% (Ravier, 1977). In another design (horizontal-stud Söderberg cells), the studs in the anode enter at the side. The fumes are removed by a hood over the whole cell (see Figure 2d). Collection efficiencies of up to 95% are obtainable today.

The potmen perform duties similar to their counterparts working on prebake pots. The crust is broken using an air hammer (crustbreaker) transported on a dolly. The tapping crew siphons the molten metal in the same way as in the prebake plant.

As the anode is consumed and lowered, the studs embedded in the anode must be removed and relocated before they make contact with the molten bath. Most of the other personnel (such as rod raisers and stud pullers) who work in the Söderberg potrooms are concerned with this task. The vertical-stud design requires that workers (stud setters) stand on a catwalk on top of the anode while adjusting bolts which attach the studs to the overhead buss bar. They can be exposed to fluorides, pitch volatiles, sulphur dioxide and possibly some carbon monoxide. For pots of the horizontal-stud design, workers (stud pullers) operate at floor level using dollies that are equipped to insert and remove the studs. Other workers in the potrooms of a horizontal-stud Söderberg facility include anode men, who install and remove the metal forms that fit around the studs and adjust the anode height.

Work around pots involves exposure to heat, and potroom workers do not work continuously in the potroom but retire to cooler areas between periods of work on the pots. Exposures are therefore intermittent.

(iii) *Pot-lining*

When the cathodes in the various cell types have served their useful lives (approximately 3-5 years), the pots are relined. The old cathode and crusts of solidified bath are removed using pneumatic hammers. The new cathode is made using a paste which is the same as or similar to that used to make the anodes and is made in a variety of shapes to line the bottom of the pots. The cathode may be rebuilt *in situ* or the old material removed and the pot relined in a special work area.

Potential exposures during the removal of the used cathode and slag materials include: dusts, if crusts are broken out dry using pneumatic drills; ammonia from nitrides in the cathode, if water is used to reduce dust production; asbestos when asbestos back-up blocks are used in pot construction; alumina, silica or other abrasives used in cleaning the pot; fumes associated with mild-steel welding; and fluorides resulting from the welding of cracks in shell repair (International Primary Aluminium Institute, 1982).

Exposures during the rebuilding of the cathode depend on the method used. When the cathode is reconstructed using a ramming paste (a mixture of sized anthracite and approximately 13% of coal-tar pitch or petroleum-pitch binder), the mixture is heated to approximately 135°C and rammed or vibrated into place in the pot. The cathode is converted to carbon in the potroom by putting it into the circuit, resulting in the emission of some volatiles. When prebaked carbon blocks are used, they are joined together with a warm, liquid pitch paste, resulting in some exposure of workers to pitch volatiles. In some instances solvents rather than heat, are used to make the paste workable (International Primary Aluminium Institute, 1982).

(iv) *Potroom maintenance*

Other workers are involved in operations in the reduction plant. These include: electricians; stud cleaners who recondition anode studs; mechanical maintenance crews who keep the crust-breaking, rod-raising and crane equipment, as well as the ventilation and exhaust systems, in working order; and crane operators, who transport the large crucibles containing the siphoned aluminium to the casting area.

Maintenance employees working in the operating area are exposed not only to the same contaminants as operating employees, but may also be exposed to higher concentrations as well as to other contaminants associated with their trades. Conducting bars of aluminium or copper are often cleaned, which may give rise to metal dusts and fumes. Gas-metal arc-welding of aluminium gives rise to ozone, among other contaminants. Employees who clean ducts may be exposed to dusts (fluorides, alumina), gases (carbon monoxide, sulphur dioxide) and pitch volatiles (International Primary Aluminium Institute, 1982).

(*b*) *Casting*

Aluminium from the pots is carried in large crucibles by an overhead crane to holding furnaces and then poured into moulds to produce aluminium ingots. Although this process is relatively simple, a number of other activities can be involved, such as alloying (in which several metals can be used), fluxing and skimming. The quality of the aluminium in the holding furnaces is checked by analysis prior to being cast into ingots. The molten metal transfer troughs and insert liners for the casting units, pouring tables and floats used to control metal flow are often made at the plant; and the furnaces must be relined periodically.

3. Exposures in the Workplace

3.1 Introduction

It should be noted that, in this section, broad generalizations are made. In practice, the level of exposure of a worker to a given contaminant is likely to vary widely between different plants, even when they operate the same types of cell.

In this description of exposures, the areas considered to be of particular interest are those associated with the aluminium reduction process, and emphasis is placed on the qualitative and quantitative exposures of workers engaged in potroom and anode preparation work. However, some potroom employees may spend some of their working life in non-potroom operations, which may also involve exposure to a wide variety of chemical and physical agents. In some plants where secondary industries (e.g., fabrication) are located on the same site, there is potential for other exposures.

The nature and level of contaminants and other agents in the potroom may be influenced by the type of pot (prebake or Söderberg) and design (vertical- or horizontal-stud Söderberg); hooding and hood exhaust rate; building ventilation; size of operation; and electrical current used (cell amperage). The actual exposure of a worker also depends on work practices. In recent years, much has been and is still being done to reduce exposures to dusts and fluoride through design of new equipment to, e.g., break crusts and siphon aluminium, and by use of respirators, air-supplied hoods, etc. Today, in some countries, major attention is given to employee protection through improved cell exhaust systems, use of microenvironments (e.g., air-purified mobile enclosures) and compulsory use of personal protective equipment.

Some of the substances that employees might work with or be exposed to in an aluminium production plant are listed in Table 2. While there may be skin contact with various materials, the main route of exposure is by inhalation.

Workers are also exposed to physical stresses, such as heat, noise and electric and magnetic fields; some may also be exposed to vibration during use of pneumatic tools.

Table 2. Airborne substances (and classes of substances) found in aluminium reduction plants[a]

Material[b]	Principal uses or sources of emission
Common airborne contaminants	
Alumina (Al_2O_3, aluminium oxide)	Potroom operations and raw materials handling
Aluminium fluoride	Potroom operations and raw materials handling
Aluminium metal	Dust generated during sanding and cutting of buss bars and spray coating of anodes to prevent anode burn
Carbon monoxide	By-product of aluminium reduction processes, furnaces, internal combustion engines
Chlorine and hydrogen chloride	Fluxing in casting operations
Coke and calcined coke	Green-carbon production, raw materials handling
Cryolite	Potroom operations, pot relining, dross recovery
Fluorides*	Potroom operations, pot repair, fluxes in casting metal treatment, welding fluxes, dross recovery
Hydrogen fluoride	Electrolytic cells
Oil mists*	Lubricating oils
Pitch - coal-tar, petroleum	Electrode production
Polynuclear aromatic compounds*	Present in binders for electrodes, potrooms, carbon and paste plants, pot relining
Sulphur dioxide	Oxidation of coke, coal, anodes, oil and gas containing sulphur
Less common airborne contaminants	
Ammonia	Low levels may be generated by reaction of water with dross or spent potlining.
Asbestos*	May be present in electrical insulators between pot anodes and cathodes or ground, thermocouple wire, asbestos cement board, spacers in casting furnaces
Copper and copper oxide	Anode rods and flux, electrical connectors, metal alloying, brazing, sanding rods and connectors
Cyanides	Potrooms, electrode production, pot soaking, cryolite recovery
Ozone	Welding
Phosphine	Gaseous by-product of dross and water
Silica	Clay refractories, sand blasting
Sodium hydroxide	Pot soaking, siphon cleaning
Welding fumes	Maintenance activities, particularly pot repair

[a]This list includes chemicals or classes of chemicals used in or formed during aluminium refining and the processes during which they are used or formed or during which exposures are most likely to occur. It was compiled from information collected during the preparation of this monograph and cannot pretend to be exhaustive.

[b]In some operations, exposures may also occur to alloying agents, such as beryllium*, cadmium*, copper, manganese and silicon.

*Chemicals (and classes of chemicals) indicated by an asterisk have been considered previously in the *IARC Monographs* series and are listed in the Appendix together with the evaluation of their carcinogenicity.

3.2 Workplace measurements

Adamiak-Ziemba *et al.* (1977) measured the concentrations of various contaminants in two Söderberg potrooms (Table 3a). In plant A the anode pins were attached to the top of the anode (vertical-stud Söderberg), and in plant B the anode pins were attached to the sides (horizontal-stud Söderberg). The cells in both plants were enclosed, but only in plant A was there a general ventilation system. Differences were found between the concentrations determined by area sampling and personal sampling. It was also evident that the exposure of the crane operator working in the potroom and exposed to contaminants from the pots was considerably higher in the plant without a general ventilation system (Table 3b).

Table 3a. Concentrations of contaminants (mg/m³, unless otherwise indicated) in two Söderberg potrooms[a]

Substance	Area samples Plant A[b]	Plant B[b]	Personal samples Plant A[c]	Plant B[d]
Hydrogen fluoride	0.36	0.32	(1.4-2.5)[e]	--
Particulate fluoride (as HF)	0.5	0.5	(1.3-1.6)	--
'Aerosols'	6.4	2.1	38	45
Alumina	5.7	1.3	2	6.1
Pitch volatiles	0.49	0.65	18	18
Benzo[*a*]pyrene	1.85 (μg/m³)	3.7 (μg/m³)	38 (μg/m³)	36 (μg/m³)
Carbon monoxide	--	--	(100-250)	--

[a]From Adamiak-Ziemba *et al.* (1977)

[b]Descriptions of the plants are given in the text; geometric mean concentrations are given.

[c]Jobs (Plant A): tapping, potmen

[d]Jobs (Plant B): 'carbon [stud] setters', potmen

[e]Where the number of samples was small, ranges are given in brackets.

Table 3b. Concentrations of contaminants (mg/m³, unless otherwise indicated) in two Söderberg potrooms (personal sampling, crane operators)[a]

Substance	Plant A	Plant B
'Aerosols'	17	62-255
Alumina	1	6.5-148
Pitch volatiles	5.2	34-107
Benzo[*a*]pyrene	13 (μg/m³)	180-800 (μg/m³)

[a]From Adamiak-Ziemba *et al.* (1977)

(*a*) *Respirable dust, alumina and aluminium*

Tedder and Chaschin (1975) noted that the size distribution of the alumina used in the electrolytic process influenced the measured concentration of particulates in potroom air. Casula *et al.* (1981) provided information supporting such size distribution differences in the potroom. They reported total dust concentrations of 0.22-2.03 mg/m³ in the side corridors of potrooms and 2.35-6.95 mg/m³ between pots. Respirable dust concentrations for the same areas ranged from not detectable to 1.12 mg/m³ and not detectable to 1.76 mg/m³, respectively. The total aluminium concentration (based on total dust) ranged from not detectable to 1.7 mg/m^3 and not detectable to 2.85 mg/m^3, respectively. In one recent study (Sjögren *et al.*, 1983), it was shown that the concentrations of aluminium in the blood of workers in the electrolytic production of aluminium were not higher than those of non-exposed referents. Concentrations of aluminium in urine (up to 25 μg/l) were significantly higher than those of the referents (median, 4 μg/l).

(*b*) *Fluorides*

(i) *Emissions*

Ravier (1977) summarized the data on emissions (kg/tonne of aluminium produced) from various types of pots. Daily fluoride emissions from a 140-kA pot (yielding about 1 tonne of aluminium per day) were 18, 14 and 8 kg of gaseous fluoride (hydrogen fluoride) per tonne of aluminium (i.e,. per day) produced for vertical-stud Söderberg, horizontal-stud Söderberg and prebake pots, respectively. Emissions of particulate fluoride were 2, 6 and 8 kg per tonne (i.e., per day), respectively.

(ii) *Concentrations in the workplace*

Some historical data on exposures to fluoride are available, and concentrations in occupational environments were summarized by the IARC (1982). Permissible levels, as fluorine, of 1-2.5 mg/m^3 have been established in many countries (International Labour Office, 1980). Kaltreider *et al.* (1972) reported 8-hour, time-weighted average exposures to fluoride for pot tenders of 2.4-3.0 mg/m^3; for tapper-carbon changers, 3.0-4.0 mg/m^3; and for crane operators, 4.0-6.0 mg/m^3 in a prebake plant in Niagara Falls, USA, from 1945-1946. The cells were not hooded. The percentages of fluoride in the gaseous state ranged from 36 in pot tending to 50 in the other two operations. Reported airborne fluoride levels in various prebake plants ranged from 0.05-0.3 mg/m^3 (Johnson *et al.*, 1973; Jahr *et al.*, 1974; Dinman *et al.*, 1976a,b,c,d), and those in Söderberg plants ranged from 0.1-3.4 mg/m^3 (Agate *et al.*, 1949; Jahr *et al.*, 1974; Gylseth & Jahr, 1975).

(iii) *Fluorides in urine*

Fluoride exposure in the aluminium industry, as expressed by urinary excretion, has been studied extensively in recent years. One summary is given by the IARC (1982). Dinman *et al.* (1976c) reported a study of over 52 000 urinary fluoride determinations over a five-year period. It was found that pre-shift urinary fluoride concentrations increased over the long term, suggesting that there is a slow uptake of fluoride by osseous tissues. The post-shift analyses correlated with changes in work practice and exposure. Pre-shift urinary concentrations in the period 1971-1974 ranged from 1-3 mg/l among anode changers, with post-shift concentrations of 3-11 mg/l. Dinman *et al.* (1976a) pointed out that because the particle size distribution of fluorides was important, the level of fluoride intake might vary with the source of alumina (on which hydrogen fluoride adsorbs) and might differ between the Söderberg plants and between the prebake plants that they were studying.

(*c*) *Polynuclear aromatic compounds*

In this section an attempt is made to provide some estimate of the tar and polynuclear aromatic hydrocarbon (PAH) exposures of employees as reported in the literature (for concentrations of benzo[*a*]pyrene, see Table 4). The composition of the airborne contaminants is complex. About 100 polynuclear aromatic compounds (PACs) have been identified by gas chromatography-mass spectrometry in an aluminium production plant (Bjørseth & Eklund, 1979). This complexity has caused problems in evaluating the industrial hygiene of aluminium production plants (see below).

Table 4. Concentrations of airborne benzo[*a*]pyrene in aluminium reduction plants

Reference	Occupation or site	Concentration (μg/m³)	Comments
Kreyberg (1959) Norway	Aluminium production (Söderberg)	0.18	1 sample, 7 polynuclear aromatic hydrocarbons quantitated in air
Konstantinov & Kuz'minykh (1971) USSR	Electrolytic shop, overhead current supply (vertical-study Söderberg), cooled anodes		124 stationary samples
	- aisles	0.6-2.2	
	- pot operator during pot processing	9.1-10.4	
	- anode operator during 'butt [stud] replacement'	28.2-230	
	- crane operator	3.6-19.1	
	Electrolytic shop, overhead current supply (vertical-study Söderberg), uncooled anodes		
	- aisles	2.4-9.4	
	- pot operator during pot processing	17.9-29.4	
	- anode operator during 'butt [stud] replacement'	221-975	
	- crane operator	10.9-73	
	Electrolytic shop, lateral current supply (horizontal-stud Söderberg)		
	- aisles	29.2-56	
	- pot operator during pot processing	44.5-245	
	- anode operator during 'butt [stud] replacement'	89-111	
	- crane operator	42.5-148	
	Electrolytic shop, prebaked anodes		
	- aisles	Not detected	
	- pot operator during pot processing	Not detected	
	- crane operator	Not detected	
Shuler & Bierbaum (1974) USA	Söderberg potroom		12 samples, 15 polynuclear aromatic compounds quantitated in air
	- tapping craneman	3.1	
	- pin setter	53.0	
	- tapper	0.2-0.3	
	- pin craneman	28.6	
	- other	0.02-0.7	
	Prebake potroom	0.03-0.1	
Adamiak-Ziemba *et al.* (1977) Poland	Söderberg potrooms		182 stationary samples, 38 personal samples
	Potroom A	1.55-2.2 (GM[a] 1.85)	
	Potroom B	3.2-4.3 (GM 3.7)	
	Potroom workers (plant A)	13-116 (GM 38)	
	Potroom workers (plant B)	7-184 (GM 36)	
	Crane operators (plant A)	5.6-30 (GM 13)	
	Crane operators (plant B)	180-800	
	Anode making	0.19-5.2	
Bjørseth *et al.* (1978) Norway	Anode plant	0.03-0.3	21 stationary samples, 7 personal samples, 34 polynuclear aromatic compounds quantitated in air
	Anode makers	0.80-27.9	
	Potroom (vertical-stud Söderberg)	0.7-9.0	
	Potroom workers (vertical-stud Söderberg)	3.4-116.3	
	Potroom (prebake anodes)	0.02-0.05	
Lindstedt & Sollenberg (1982) Sweden	Potroom (Söderberg)	1.8-5.3	
	Pin setters	10-24	
	Flex raisers	23-51	
	Anode service	4-22	
	Cathode repair	2-17	

[a]GM, geometric mean

Kreyberg (1959), in an early study of the content of airborne contaminants in industrial settings, reported the following analyses of airborne samples of tars from an aluminium smelter (Söderberg process; μg/m³): fluoranthene, 1.0; benz[*a*]anthracene, 0.80; phenanthrene, 0.27; benzo[*a*]pyrene, 0.18; benzo[*e*]pyrene, 0.10; pyrene, 0.69; benzo[*ghi*]perylene, 0.05; anthracene, 'present'; and trace amounts of anthanthrene and coronene.

Konstantinov and Kuz'minykh (1971) studied the concentrations of tarry substances and of benzo[*a*]pyrene (using fluorescent analytical methods for both) in four aluminium production facilities. In general, the concentrations of benzo[*a*]pyrene in the prebake potrooms were found to be lower than in the Söderberg potrooms. In potrooms containing Söderberg pots with an overhead current supply (presumably vertical-stud) with cooled or uncooled anodes, the highest levels for the operator were found during 'butt [stud] replacement'. In a facility with potrooms with lateral current supply pots (presumably horizontal-stud), general potroom worksite benzo[*a*]pyrene concentrations were higher, although levels during 'butt [stud] replacement' appeared to be slightly lower than at the other two facilities. At the facility using 'preburned electrodes' (presumably prebake), benzo[*a*]pyrene concentrations were below the level of detection. In general, the concentrations of tarry substances at these facilities paralleled the benzo[*a*]pyrene levels.

Shuler and Bierbaum (1974) reported pitch-volatile concentrations (benzene soluble) at three aluminium production facilities. Limited analyses for some individual PACs were also conducted. In general, the results of this study are consistent with the results of Konstantinov and Kuz'minykh (1971). The highest pitch-volatile levels were found in potrooms with horizontal-stud Söderberg pots, followed closely by the concentrations in potrooms where vertical-stud Söderberg pots were used (Table 5). The lowest concentrations were found in prebake potrooms. Pitch-volatile concentrations were generally lower in carbon plant areas than in the

Table 5. Pitch-volatile concentrations (benzene soluble) at aluminium reduction facilities[a]

	No. of samples	Pitch volatile levels, range (mg/m³)[b]	TWA[c] (mg/m³)
Prebake potroom			
All workers	8	ND-2.1	0.4
Vertical-stud Söderberg potroom			
Potman	10	ND-6.8	2.2
Tapper	5	1.0-63.4	12.0
Head stud setter	3	11.3-19.0	7.5
Other workers	12	ND-8.1	2.6
Horizontal-stud Söderberg potroom			
Potman	4	4.6-8.7	4.5
Tapper	2	3.0-3.6	2.0
Stud driver	4	8.3-26.9	12.0
Other workers	15	2.2-60.0	7.2
Carbon plant Prebake anode-block production			
All workers	17	ND-16.7	2.7
Carbon plant Prebake anode-paste production			
All workers	10	ND-12.6	3.4

[a]Compiled from Shuler and Bierbaum (1974)

[b]ND, not detected

[c]TWA, mean time-weighted concentration for the job group

Söderberg potrooms. This may reflect the relatively lower temperatures at which the pitch was usually handled. A limited analysis for PACs at one facility (vertical-stud Söderberg potroom) demonstrated the presence of several individual PACs in some of the personal samples with higher pitch-volatile concentrations. The PAHs most commonly found included chrysene (not detected-82.7 $\mu g/m^3$); benzo[*b*]fluoranthene (not detected-58.2 $\mu g/m^3$); benzo[*a*]pyrene (0.03-53.0 $\mu g/m^3$); and benz[*a*]anthracene (not detected-42.1 $\mu g/m^3$).

Steinegger (1981), in a study in progress, reported levels of cyclohexane-soluble matter from workplace samples taken in aluminium production facilities. Concentrations were highest in the paste plant (0.12-0.33 mg/m^3) and potrooms (0.16-0.24 mg/m^3). Levels of 0.07-0.1 mg/m^3 were found in baking areas.

Bjørseth *et al.* (1978) reported on the profile of PACs in Söderberg and prebake potrooms (closed anode) (Table 6). Gas chromatographic analyses were conducted on particulate matter collected on a filter and on vapours absorbed into ethanol cooled by dry ice. It is clear that the concentrations of PACs were considerably less in prebake than in Söderberg operations, especially as the prebake samples analysed were those with sufficient quantities to analyse and thus reflected higher than average concentrations. The Söderberg plant also showed a different PAC profile than the prebake plant. A significantly higher fraction of the PACs in the Söderberg samples belonged to the higher-boiling PAC compounds than in the prebake anode facility. This difference reflects the differences in the two aluminium reduction processes.

Table 6. Levels of polynuclear aromatic compounds ($\mu g/m^3$) in the atmospheres of areas of an aluminium production plant[a]

Polynuclear aromatic compound	Vertical-stud Söderberg plant (range of 10 samples)		Prebake plant (range of 6 samples)	
	Particulate	Gaseous	Particulate	Gaseous
Naphthalene	ND-4.0	0.72-311.3	--	ND-12.04
Methylnaphthalene	ND-0.12	1.22-123.5	--	ND-22.3
Biphenyl	ND-0.42	5.38-381.0	--	ND-14.48
Dibenzofuran	ND-0.46	2.09-61.0	--	ND-3.57
Fluorene*	ND-0.53	6.31-135.2	--	ND-8.85
9-Methylfluorene	ND-0.1	0.12-11.0	--	ND-0.28
2-Methylfluorene	ND-0.34	0.85-13.0	--	ND-0.86
1-Methylfluorene	ND-0.23	0.27-13.0	--	ND-1.1
Dibenzothiophene	0.2-1.2	3.46-58.0	--	ND-5.2
Phenanthrene*	ND-12.0	39.9-446	ND-0.16	1.91-49.0
Anthracene*	ND-1.8	2.98-32.6	--	ND-5.27
Carbazole*	ND-2.8	ND-1.4	--	ND-0.20
Methylphenanthrene* and methylanthracene	ND-3.2	0.17-15.3	--	ND-1.5
Fluoranthene*	5.5-30.0	14.1-162	?0.14-0.66	0.88-13.4
Dihydrobenzo[*a*]- and dihydrobenzo[*b*]fluorene	ND-0.91	ND-1.8	--	ND-0.20
Pyrene*	3.7-24.8	8.03-91.1	0.08-0.37	0.40-7.84
Benzo[*a*]fluorene*	1.4-9.1	ND-7.5	--	ND-0.70
Benzo[*b*]fluorene*	0.9-5.7	ND-2.3	--	ND-0.15
1-Methylpyrene	0.3-3.0	ND-0.16	--	--
Benzo[*c*]phenanthrene*	0.4-7.9	ND-0.36	--	--
Benz[*a*]anthracene	1.0-15.0	ND-0.21	0.05-0.26	--
Chrysene/triphenylene*	2.1-30.1	ND-0.44	0.07-0.36	--

Polynuclear aromatic compound	Vertical-stud Söderberg plant (range of 10 samples)		Prebake plant (range of 6 samples)	
	Particulate	Gaseous	Particulate	Gaseous
Benzo[*b*]- and benzo[*k*]fluoranthene*	1.1-26.9	--	ND-0.28	--
Benzo[*e*]pyrene*	0.8-12.2	--	ND-0.09	--
Benzo[*a*]pyrene	0.7-9.0	--	ND-0.05	--
Perylene*	ND-2.4	--	--	--
o-Phenylenepyrene	0.4-5.3	--	ND-0.16	--
Benzo[*ghi*]perylene*	0.4-5.8	--	ND-0.32	--
Anthanthrene*	ND-0.32	--	--	--
Coronene*	ND-1.2	--	--	--
Dibenzopyrene*	ND-1.2	--	--	--
Total polynuclear aromatic compounds	19.7-202	86.0-1564.1	0.56-1.7	4.56-123.95

[a]From Bjørseth *et al.* (1978). Compounds marked with an asterisk (*) were considered in Volume 32 of the *Monographs* (IARC, 1983). ND, not detected.

It is also interesting to note that the PAC profile of air samples taken from the topside of a coke oven was similar to that for the Söderberg plant (Bjørseth & Eklund, 1979). This similarity reflects the similar operating temperatures of the two processes.

Bjørseth *et al.* (1978) also sampled in an anode fabrication plant (carbon plant). Total PAC concentrations in that facility ranged from 3.4 to 73.3 μg/m³ on particulates and 10.5 to 262.0 μg/m³ in the gaseous phase (five samples). The PAC profile of the baking section of the anode plant was similar to that of the prebake plant, reflecting the similar materials and processes. The results of personal sampling in all three facilities are shown in Table 7 (particulates only). There was considerable variation in exposure even among persons doing the same job, and levels determined by personal monitoring differed substantially from those determined by stationary sampling.

Table 7. Typical combined total concentrations of approximately 40 polynuclear aromatic compounds determined by personal sampling[a]

Site and occupation	Concentration (μg/m³)
Anode plant	
Coke packer	36.84
Pitch-bin worker	373.7
Operator	10.58
Pitch-dust sweeper	10.82
Prebake plant	
Tapper	0.87
'Coke puller' [Carbon setter]	2.0
'Coke puller' [Carbon setter]	0.87
Crust breaker	0.52
Söderberg plant	
Tapper	37.7
Pin [stud] puller	2790.5
Pin [stud] puller	926.8

[a] Summarized from Bjørseth *et al.* (1978). The sampling period varied from approximately 2-6 hours

Bjørseth *et al.* (1981) also provided a more detailed analysis of individual PACs associated with various jobs (sampling for PACs adsorbed on particulates). These results confirmed his previous finding that stud-pullers had the highest exposures in Söderberg plants. The results also demonstrated significant variability in individual exposures even within the same job type. However, in all personal samples, total PAC exposures were greater than 40 $\mu g/m^3$.

Recently Becher and Bjørseth (1983) evaluated PAH exposure by an analysis of the metabolites and aromatic hydrocarbons found in urine. Among non-smokers, the average urinary PAH levels were approximately six times higher in the exposed group than in the non-exposed group. No significant difference in PAH levels was found among exposed and non-exposed smokers. In the non-exposed group, smokers had significantly higher PAH levels than non-smokers, but no significant effect from smoking was found in the exposed group. The authors also noted the small amounts of PAHs found in the urine of aluminium workers (1-9 µg per mmol of creatinine), even though the workers were exposed to 1000 times the PAH levels in the urban atmosphere.

4. Biological Data Relevant to the Evaluation of Carcinogenic Risk to Humans

The studies of carcinogenicity and mutagenicity reported in this section include only those on samples of complex mixtures taken from the industry.

4.1 Carcinogenicity studies in experimental animals

Skin application

Mouse: Samples of airborne particulate polynuclear organic matter collected from two sites in an aluminium production plant, one being the air-control system of an anode-paste plant and the other air in an unspecified area collected on electrostatic precipitator plates, were tested for carcinogenicity by topical applications of 50 mg in toluene (1:1) twice weekly to the skin of C3H mice. Control groups were treated with benzo[*a*]pyrene in toluene or with toluene alone. Tumours of the skin were produced by the first sample in toluene (containing 0.11% benzo[*a*]pyrene) in 17/18 mice and were malignant in 15; the second sample in toluene (containing 0.62% benzo[*a*]pyrene) produced malignant skin tumours in 15/17 mice. The average time to appearance of the first tumours was 24 weeks for the first sample and 18 weeks for the second sample. No tumour was produced in toluene-treated controls. Positive controls developed papillomas and carcinomas of the skin. Malignant tumours of the skin occurred in 9/16, 24/39 and 14/16 mice treated with 0.06%, 0.1% or 0.3% benzo[*a*]pyrene in toluene, with average times to appearance of the first tumours of 38, 32 and 25 weeks, respectively (Bingham *et al.*, 1979).

4.2 Other relevant biological data

(*a*) *Experimental systems*

No data were available to the Working Group on toxic or reproductive effects.

Tests for genetic and related effects in experimental systems

Emission samples taken from the air in an aluminium smelting facility (two production plants using Söderberg electrodes and an anode-paste plant) were collected at 40-50°C on glass-fibre filters and an XAD-2 adsorbant and tested for mutagenicity in *Salmonella typhimurium* strain TA100 in the presence and in strain TA98 in the presence and absence of an Aroclor-induced rat-liver metabolic system (S9). Smelting plant 1 was an older plant with horizontal studs, and smelting plant 2 used vertical studs. All of the samples taken from the two smelting plants were mutagenic to strains TA98 and TA100 in the presence of S9; some samples were also mutagenic to strain TA98 in the absence of S9. Air samples taken at the mixer and pitch melter in the anode paste plant were mutagenic to strains TA98 and TA100 only in the presence of metabolic activation. The samples from plant 1 included the ventilation air from one potroom and emissions from the pot gas collected after the air had passed through four scrubbers. The samples from plant 2 included separate samples of ventilation air collected after passing through either wood grid or floating bed scrubbers. The ventilation-air samples from plant 2 were taken from potrooms using electrodes containing a normal pitch content and from potrooms using electrodes containing 40% less pitch. A substantial reduction (62-85%) in the mutagenic activity of the air (after scrubbing) was observed when the normal-pitch paste was replaced with low-pitch paste (Alfheim & Wikström, 1984).

Air particle samples (cyclohexane extracted) taken with filters in an aluminium Söderberg production facility in the anode-paste plant and potroom were both mutagenic to *S. typhimurium* strain TA98 in the presence and absence of metabolic activation with S9 and to strain TA100 with S9 (Krøkje *et al.*, 1984a).

(*b*) *Effects in humans (other than cancer)*

Toxic effects

Respiratory effects. According to a review by Dinmam (1977), the occurrence of bronchial asthma in potroom workers was first reported in Norway by Frostad (1936). Evang (1938) found that 18 of 190 potroom workers suffered from bronchial asthma compared with 32 patients in a local population of about 3800. Subsequently, bronchial asthma as well as dyspnoea, wheezing and coughing have been observed in potroom workers in various countries (Bruusgaard, 1960; Midttun, 1960; Field & Milne, 1975; van Voorhout, 1977; Šaric *et al.*, 1979; Maestrelli *et al.*, 1981; Vale, 1981). Smith (1977) also observed bronchial asthma and other persistent respiratory symptoms among workers in prebake potrooms.

For men with 20 years' employment in the prebake process, the standardized mortality ratio (SMR) for asthma was 260.2, based on seven observed deaths ($p < 0.01$). Potroom workers and carbon plant workers in the Söderberg process had excess mortality from emphysema (SMR, 194.0, $p < 0.05$; and SMR, 267.1, $p < 0.05$, respectively) (Rockette & Arena, 1983).

Complex mixtures found in potrooms can include compounds that have been reported individually to cause damage or irritation to the respiratory tract: hydrogen fluoride gas (Machle *et al.*, 1934; Dieffenbacher & Thompson, 1962; Mayer & Guelich, 1963; Ricca, 1966, 1970) and sulphur dioxide (Charan *et al.*, 1979).

Mylius and Gullvåg (1984) found a significant increase in the number of macrophages in sputum expectorates from 84 potroom workers over that in 36 controls (teachers and students). Similar results were obtained at a second facility, where workers in the potrooms (smokers) showed higher macrophage counts than control smokers. A more than additive effect of smoking and exposure was observed.

Chronic pulmonary disease. In a historical prospective (cohort) study of 2103 workers in a prebake aluminium smelter, an increased number of cases of emphysema was observed (SMR, 204) (Milham, 1979).

Clonfero *et al.* (1978) examined 197 potroom workers in areas where the mean values of dust and fluoride were below the threshold limit value of 2.5 mg/m^3. Chronic bronchitis was graded by questionnaire into three categories: there was an increasing prevalence of each grade with years of exposure. Respiratory function tests showed that forced expiratory volume, as a percentage of the predicted value, diminished with length of exposure, an effect that was unrelated to cigarette smoking.

Discher and Breitenstein (1976) studied chronic pulmonary disease in 457 male potroom workers and former workers who were employed during one year at either of two plants using a prebake process (274 workers) or at one plant using the Söderberg process (183 workers), compared with 228 skilled workers not currently exposed to similar occupational risks. The controls were matched to potroom workers by age and smoking history. A questionnaire on occupational history, smoking habits and respiratory symptoms was administered and spirometry was performed; there was no response for 29% of cases and 14% of controls. Chronic respiratory disease was defined as a combination of one or more major symptoms (coughing, wheezing and shortness of breath) given in the questionnaire, plus abnormal spirometry. No significant difference between the exposed workers and the controls was observed. However, this was a survivor population with a high turnover - of 85 new hires in one year, 36% were no longer in the potrooms one year later, and, of these, 75% were under 36 years of age. The authors suggested possible self-selection out of the potroom because of respiratory problems.

Fluorosis. Inhalation of fluoride poses a potential hazard in workers in primary aluminium production, but the majority of workers are clinically unaffected. Clinical fluorosis, which is rare, commences with stiffness in the lower back followed by pain and then limitation of rotation of the trunk. Later, the spine becomes rigid and stiff, with restriction of chest movement and of the large joints, particularly the hip, that is accompanied by osteosclerosis.

Fluorosis was reported first by Møller and Gudjonsson (1932) and investigated by Roholm (1937). Later, Hodge and Smith (1977) reviewed the literature on occupational fluoride exposure, and their main conclusion was that osteosclerosis develops when air fluoride concentrations are above 2.5 mg/m^3 and urinary fluorides levels exceed 9 mg/l for prolonged periods.

Dermatological and vascular effects. Bazyka *et al.* (1977) studied workers in the electrolysis shop of an aluminium plant and noted vascular changes in the skin (telangiectasia). The incidence of these changes increased with duration of exposure. Capillary changes were also seen in the conjunctiva. Thériault *et al.* (1980) studied aluminium workers in a primary smelter and found that the number of telangiectatic lesions increased in size and dispersion with time of exposure. There was no significant variation in numbers of skin lesions in men who did not work in the potroom.

Dermatitis has been observed in potroom workers (Midttun, 1960).

Cardiovascular effects. An SMR of 160.1 ($p < 0.05$) for the category 'all other heart disease' was observed among white workers in a plant using the Söderberg process. The excess in this non-specific remainder category is difficult to interpret. The SMR for heart disease was 119.1 ($p < 0.01$) and both whites and non-whites showed the excess (SMR, 118.9, $p < 0.01$;

SMR, 124.5, respectively). Among several subcategories, the only statistically significant excess was for ischaemic heart disease for both whites and non-whites (SMR, 125.5, $p < 0.01$, and 217.0, $p < 0.01$, respectively) (Rockette & Arena, 1983).

Mutagenicity and chromosomal effects

Sputum samples from aluminium refinery (Söderberg process) workers in the potrooms and the anode-paste plant were taken and divided according to smoking habits. The pooled sputum samples from potroom workers were assayed in *Salmonella typhimurium* strains TA98 and TA100 with metabolic activation at only one dose. Sputum from potroom workers, both smokers and non-smokers, was mutagenic, and the number of net revertants was reported to be significantly higher in the potroom smokers, while sputum samples from the controls (non-exposed) were negative whether they were smokers or non-smokers (Krøkje *et al.*, 1984b). [The significance of the comparative mutagenicity of the exposed smokers and non-smokers is difficult to evaluate in the absence of dose-reponse data.]

Haugen *et al.* (1983) reported, in an abstract, that there was no significant increase of sister chromatid exchange in peripheral lymphocytes in exposed workers [number not specified] at an aluminium plant [process not specified] as compared with matched controls, who were office workers. Smokers in both groups had a significantly higher rate of sister chromatid exchange than non-smokers. The total amount of polynuclear aromatic hydrocarbons (PAHs) in the working atmosphere varied from 52 to 268 $\mu g/m^3$, with a mean of 126 $\mu g/m^3$.

4.3 Case reports and epidemiological studies of carcinogenicity in humans

Konstantinov and Kuz'minykh (1971) compared cancer mortality rates in two aluminium production plants in the USSR, one using the Söderberg process, the other prebake anodes. Expected figures were computed from regional rates. Excesses of all cancers and of lung cancer were claimed for the Söderberg-process workers over the ten-year period 1956-1966; an increased incidence of skin cancer was reported also, particularly in young workers. [No raw data were given.]

Konstantinov *et al.* (1974) investigated mortality from cancer among potroom workers in three aluminium plants, two using the Söderberg process and the other using prebake anodes. Mortality from cancer in the plants was compared with that of the general population of the cities and provinces in which the aluminium plants were located. Elevated ratios were reported for lung cancer in the two plants using the Söderberg process and for skin cancer in one. [The absence of information both on the study population and the reference population precludes evaluation of these results. It was not possible to establish whether any of the factories included in the present study had already been surveyed in the earlier report by the same author.]

Milham (1976) noted an elevated proportionate mortality ratio (PMR) in aluminium workers in Washington State, USA, for the period 1950-1971 for all neoplasms, and for cancer of the pancreas (PMR, 204) and for lymphoma (PMR, 250) among potroom workers.

Milham (1979) used a historical prospective (cohort) study to estimate standardized mortality ratio (SMR) for all causes of death in 2103 aluminium production workers in a prebake plant. The cohort consisted of all men who had worked in the plant for three or more years, including at least one year between 1946 and 1962. Expected figures were computed on the basis of national rates. Smoking histories were not obtained. The SMRs were 117 for lung cancer (based on 35 cases), 180 for pancreatic cancer (based on nine cases) and 184 for lymphatic and haematopoietic cancer (based on 17 cases, $p < 0.05$). The group was divided into exposed and non-exposed workers: exposure was defined as occurring in carbon plants, rodding, potlining, potrooms and quality control; 'unexposed' workers were mechanics, maintenance workers, electrical workers, yard workers, metal products workers, guards, janitors, storeroom workers, south plant workers, exempt workers, welders, carpenters, pipe shop workers, and masons. There was a SMR of 643 for lymphosarcoma/reticulosarcoma among 'exposed' workers. A SMR of 391 for fatal benign tumour of the brain was based on five cases in 'unexposed' workers.

Gibbs and Horowitz (1979) studied mortality from lung cancer in 5406 men (Cohort 1) and 485 men (Cohort 2) employed in aluminium production in Canada. Both plants used the Söderberg process. Cohort 1 was followed for 24 years from 1950-1973, Cohort 2 for 23 years from 1951-1973. More than 99% of the men in Cohort 1 and just less than 98% in Cohort 2 were traced. There were 84 deaths from lung cancer in Cohort 1 and 11 in Cohort 2. For each man and each year of follow-up, three exposure variables were defined: the number of years of exposure to 'tars' [these were probably pitch fumes], the number of years since first exposure to tars, and an index of the degree of exposure, expressed in 'tar years', calculated as: 0.25 x the number of years of 'some' tar exposure + the number of years of 'definite' tar exposure. Time worked in prebake operations was classified as 'no tar exposure'. Expected deaths were computed from regional age- and sex-specific death rates. SMRs were not different whether the men had ever been exposed to tars or not. A dose-response relationship was seen between lung cancer mortality and both years of exposure (more pronounced for 21 years or more) and tar-years for the combined cohorts. The overall risk for lung cancer among the workers in Cohort 1 was significantly greater than that for the general population. [Smoking histories were not available. Because of the small size of Cohort 2, equivalent conclusions could not be drawn.]

Simonato (1981) reanalysed the data and noted that the excess of lung cancer deaths was highly significant in the subgroup with the longest exposure, the longest period since first exposure and the heaviest exposure to tar. The tar-year index appears to be the most sensitive exposure index related to excess lung cancer deaths.

In a further analysis of the same cohort, Gibbs (1983) examined the incidences of neoplasms other than those of the lung. Although the number of cases was small (10), a clear relationship emerged between years of exposure and incidence of carcinoma of the bladder, with SMRs of 36.7, 53.4, 166.8 and 919.1 in the groups of workers exposed for 0, < 1-10, 11-20 and more than 20 tar-years, respectively.

Giovanazzi and D'Andrea (1981) examined the mortality of 494 workers in a primary aluminium production plant, using mainly the Söderberg process, during the period 1965-1979. Among the 212 potroom workers there were 40 deaths (2115 person-years), whereas among the 282 workers in other departments there were 13 deaths (3191 person-years). Expected figures were computed on the basis of national mortality data and municipal data. There were significantly more observed deaths in potroom workers from all tumours than expected (14 observed, 8 expected; SMR, 175; $p < 0.05$) and for liver cirrhosis (7 observed, 1.8 expected; SMR, 389; $p < 0.01$); a statistically non-significant excess of lung cancer (SMR, 174, based on only four cases) was also observed.

Andersen *et al.* (1982) studied cancer incidence and mortality in 7410 male employees in primary aluminium production in Norway during the period 1953-1979. For cancer incidence, expected figures were computed on the basis of five-year, age-specific regional incidence; for mortality, national rates were used. The observed number of cases exceeded that expected for cancers of the lung, kidney, larynx, bladder and for leukaemia. Only for cancer of the lung was the excess statistically significant, with 57 cases observed compared to 36 expected, in the local population; however, when an expected number of 54 was computed on the basis of national rates, the significance vanished. A large excess risk was seen in those workers in the processing department of older plants (started around 1915) with more than 25 years of employment, observed cases being 2.4 times the expected; however, in process workers in new plants (started around 1950) with employment of between 1.5 and 4 years, the observed number of cases was 2.9 times that expected. [Information concerning previous occupational history was incomplete and no information on smoking habits was available. It was impossible to distinguish between those who had worked with the Söderberg and those with the prebake system, because many had worked with both.]

Wigle (1977) reported a high incidence of bladder cancer in the Chicoutimi census division of Quebec, Canada where there is aluminium production. During the five years 1969-1973, in a population of 163 350 (1971), 81 newly diagnosed cases of bladder cancer were recorded; 51 were expected on the basis of incidence rates for the Province of Quebec.

The hypothesis that this finding was related to the presence there of the aluminium production industry, which is mainly based on the Söderberg process, was tested in a case-control study (Thériault *et al.*, 1981). In the Chicoutimi census division, 96 cases of bladder cancer were diagnosed between 1970 and 1975 and identified from records of the Province Tumor Registry and the Regional Hospital. The investigation was restricted to the 81 males in the group, who were matched by age and sex with neighbours as controls. This compared to 48.5 cases expected on the basis of age-specific incidence rates for bladder cancer estimated from the Province Tumor Registry. There was no difference between cases and controls with regard to previous urinary tract disease (22:19), alcohol intake (45:46) or coffee drinking (61:60); years of regular intake of coffee and weekly intake of coffee also showed no significant difference. When smoking habits were compared, there was no significant difference between cases and controls, except in the mean number of years of smoking. The relative risk for men working in the aluminium production plant who did not smoke was 1.9, but those who smoked cigarettes had a relative risk of 5.7. [The tumours were classified only on the basis of provincial and hospital records.]

Rockette and Arena (1983) studied mortality patterns of 21 829 workers with five or more years of employment between 1946 and 1977 in 14 aluminium production plants in the USA. Three types of process were used in the plants: prebake (7 plants), vertical-stud Söderberg (1 plant) and horizontal-stud Söderberg (5 plants). In addition to studying overall mortality patterns for selected causes of death, a more detailed analysis was done for individual processes relative to years of cumulative employment. SMRs were used to compare cause-specific mortality of the workers with that of the US male population. Data from one plant (plant 3) were analysed separately, since all three processes were used, and in most cases no indication was given in the job history of the process in which a man worked. There was no statistically significant excess for lung cancer in any subgroup. A significant excess was noted for lung cancer among white workers at plant 5 where the horizontal-stud Söderberg process was used (SMR, 162.0, based on 25 observed deaths), but, if county rates were used instead of the total US rate, the resultant SMR was 86.2. A statistically significant SMR of 365.9 for lung cancer ($p < 0.05$) (with 6 observed and 1.64 expected deaths) was seen in men with 25 or more years of employment in plant 3. Further investigation showed that this excess occurred among men employed in the carbon bake department. For men employed in the potroom or carbon department of the Söderberg plants for five or more years, the SMR for bladder cancer was found to be 236.2 (95% confidence intervals, 45.9, 273.0), based on six observed deaths. This excess occurred in whites, for whom the SMR was 408 ($p < 0.05$). Among the 94 deaths from haematolymphopoietic cancers, 22 were from lymphosarcoma and reticulosarcoma, and 13 of these occurred in plants 1 and 11 (prebake) where the SMRs were 176.5 and 282.3, respectively, but were not statistically significant. For those men in plant 11 with employment in the potroom or carbon department, the SMR was 340.3 ($p < 0.05$). [The same phenomenon was noted by Milham (1979)]. The largest SMR in the haematolymphopoietic category was for leukaemia and aleukaemia (127.9) based on 43 observed deaths. For men with less than 15 years of cumulative employment in the potroom or carbon department in plants using the Söderberg process, there was a statistically significant excess (SMR, 274.7; $p < 0.05$). Although it was not statistically significant, there was a more than 20% excess of stomach cancer in white workers in both the prebake and Söderberg processes. Of the 55 observed deaths in the cohort, 22 occurred in plant 1 (prebake) (SMR, 173.9; $p < 0.05$). This excess occurred in men with potroom exposure and time since first employment of at least 30 years (SMR, 295.1; $p < 0.01$). Examination of the work histories revealed no common factor that might explain the excess. Men employed for 15 or more years in the prebake and horizontal-stud Söderberg process and in plant 3 had an increased risk of pancreatic cancer (SMR, 222.2; $p < 0.05$; 271.3; $p < 0.03$; 168.1, respectively). Because an estimated 16% of the cohort from plant 11 (prebake) was missing, it was not included in the analysis that was based on duration of employment. However, an excess of pancreatic cancer was also seen in the potrooms in this plant (SMR, 197.4), as noted by Milham in a different cohort from the same plant.

5. Summary of Data Reported and Evaluation

5.1 Exposures

Primary aluminium production plants are located in about 40 countries. The two main methods used for aluminium production are Söderberg and prebake, which encompass a number of processes and job categories. Substantial exposures to airborne polynuclear aromatic compounds have been measured in certain occupational settings in this industry. Exposures have been higher in potrooms of plants using the Söderberg process than in those using the prebake process; some workers may have been exposed to both processes. Exposures to fluorides and a variety of other contaminants also occur in potrooms.

5.2 Experimental data

Two incompletely characterized samples of airborne particulate polynuclear organic matter from an aluminium production plant were tested for carcinogenicity by skin application to mice, resulting in a high incidence and early appearance of papillomas and carcinomas of the skin.

Air samples from various locations in two aluminium production facilities were mutagenic to *Salmonella typhimurium*. No data on cell transformation were available.

Overall assessment of data from short-term tests on samples from aluminium production plants[a]

	Genetic activity			Cell transformation
	DNA damage	Mutation	Chromosomal effects	
Prokaryotes		+[b]		
Fungi/ Green plants				
Insects				
Mammalian cells (*in vitro*)				
Mammals (*in vivo*)				
Humans (*in vivo*)				
Degree of evidence in short-term tests for genetic activity: *Inadequate*				Cell transformation: No data

[a]The groups into which the table is divided and '+', '−' and '?' are defined on pp. 16-17 of the Preamble; the degrees of evidence are defined on pp. 17-18.

[b]Ventilation air from Söderberg potrooms and samples from anode-paste plants

5.3 Human data

Asthma, chronic pulmonary disease and skin lesions occur in potroom workers. Fluorosis has occurred in workers in the aluminium production industry.

The lung has been the most common site identified for excess cancer in populations of aluminium production workers. Overall, there was a borderline excess in relative risk, with some studies showing a doubling of risk and some showing no excess. No smoking history was given in any of these studies. In one study in which populations in the industry were compared on the basis of their exposures to pitch volatiles, there was a relationship between incidence of lung cancer and length of exposure, and a significant excess of cancer among workers who had worked for 21 years or more.

In three studies in the same aluminium-producing area, an increased risk of bladder cancer was associated with work in aluminium production in plants where primarily the Söderberg process was used. In one study in which smoking was controlled for, while there was a borderline excess in risk for non-smokers, the risk for smokers was markedly increased.

An excess of lymphosarcoma/reticulosarcoma was noted only in two cohort studies which covered partially the same population.

Statistically significant excess risks of pancreatic cancer and leukaemia were noted as isolated findings in two studies and in one study, respectively.

5.4 Evaluation[1]

The available epidemiological studies provide *limited evidence* that certain exposures in the aluminium production industry are carcinogenic to humans, giving rise to cancer of the lung and bladder. A possible causative agent is pitch fume. There is *inadequate evidence* that occupational exposures in the aluminium production industry result in haematolymphopoietic and pancreatic cancer.

There is *sufficient evidence* that samples of particulate polynuclear organic matter from one aluminium production plant were carcinogenic to experimental animals. However, because of the incomplete characterization of the samples tested, no evaluation of the carcinogenicity to experimental animals of complex mixtures that occur in the aluminium production industry could be made.

A number of individual polynuclear aromatic compounds for which there is *sufficient evidence* of carcinogenicity in experimental animals have been measured at high levels in air samples taken from certain areas in aluminium production plants.[2]

Taken together, the available evidence indicates that certain exposures in the aluminium production industry are probably carcinogenic to humans.

6. References

Adamiak-Ziemba, J.A., Ciosek, A. & Gromiec, J. (1977) The evaluation of exposure to harmful substances emitted in the process of the production of aluminium using selfbaking anodes (Pol.). *Med. Pr.*, *28*, 481-489

Agate, J.N., Bell, G.H., Boddie, G.F., Bowler, R.G., Buckell, M., Cheeseman, E.A., Douglas, T.H.J., Druett, H.A., Garrad, J., Hunter, D., Perry, K.M.A., Richardson, J.D. & Weir, J.B. de V. (1949) *Industrial Fluorosis. A Study of the Hazard to Man and Animals near Fort William, Scotland* (*Medical Research Council Memorandum No. 22*), London, Her Majesty's Stationery Office

ALCAN (1976) *75 Years of Canadian Aluminium Development*, Montreal, Alcan Smelters & Chemicals Ltd

ALCAN (1978) *The Aluminium Industry in Brief*, Montreal, Alcan Smelters & Chemicals Ltd

Alfheim, I. & Wikström, L. (1984) Air pollution from aluminum smelting plants. I. The emission of polycyclic aromatic hydrocarbons and of mutagens from an aluminum smelting plant using the Söderberg process. *Toxicol. environ. Chem* (in press)

American Bureau of Metal Statistics (1983) *Non-Ferrous Metal Data 1982*, Washington DC, National Technical Information Service, pp. 89-90

[1]For definitions of the italicized terms, see Preamble, pp. 15 and 19.

[2]See Appendix, Table 2

Andersen, A., Dahlberg, B.E., Magnus, K. & Wannag, A. (1982) Risk of cancer in the Norwegian aluminium industry. *Int. J. Cancer*, *29*, 295-298

Bazyka, A.P., Logunov, V.P., Selivonenko, V.G. & Kozlenko, V.V. (1977) Relationship between occupational vascular affections of the skin and occupational disorders of the cardiovascular system in workers of the electrolysis shop of an aluminium plant (Russ.). *Vestnik. Dermatol.*, *7*, 73-77

Becher, G. & Bjørseth, A. (1983) Determination of exposure to polycyclic aromatic hydrocarbons by analysis of human urine. *Cancer Lett.*, *17*, 301-311

Bingham, E., Trosset, R.P. & Warshawsky, D. (1979) Carcinogenic potential of petroleum hydrocarbons. A critical review of the literature. *J. environ. Pathol. Toxicol.*, *3*, 483-563

Bjørseth, A. & Eklund, G. (1979) Analysis for polynuclear aromatic hydrocarbons in working atmospheres by computerized gas chromatography-mass spectrometry. *Anal. chim. Acta*, *105*, 119-128

Bjørseth, A., Bjørseth, O. & Fjeldstad, P.E. (1978) Polycyclic aromatic hydrocarbons in the work atmosphere. I. Determination in an aluminum reduction plant. *Scand. J. Work Environ. Health*, *4*, 212-223

Bjørseth, A., Bjørseth, O. & Fjeldstad, P.E. (1981) Polycyclic aromatic hydrocarbons in the work atmosphere. Determination of area-specific concentrations and job-specific exposure in a vertical pin Söderberg aluminum plant. *Scand. J. Work Environ. Health*, *7*, 223-232

Bruusgaard, A. (1960) Asthma-like disease in Norwegian aluminium furnace room workers (Norw.) *Tidsskr. Nor. Laegefor.*, *17*, 796-797

Casula, D., Cherchi, P., Marraccini, L., Atzeri, S., Spinazzola, A. & Spadaccino, E. (1981) Industrial hygiene problems in alumina and aluminium production plants (Ital.). *Med. Lav.*, *4*, 283-290

Charan, N.B., Myers, C.G., Lakshminarayan, S. & Spencer, T.M. (1979) Pulmonary injuries associated with acute sulfur dioxide inhalation. *Am. Rev. resp. Dis.*, *119*, 355-360

Clonfero, E., Cortese, S., Saia, B., Marcer, G. & Crepet, M. (1978) Lung disorders in aluminium potroom workers (Ital.). *Med. Lav.*, *69*, 613-619

Dieffenbacher, P.F. & Thompson, J.H. (1962) Burns from exposure to anhydrous hydrofluoric acid. *J. occup. Med.*, *4*, 325-326

Dinman, B.D. (1977) *The respiratory condition of potroom workers: Survey of IPAI member companies: Preliminary report*. In: Hughes, J.P., ed., *Health Protection in Primary Aluminium Production*, London, International Primary Aluminium Institute, pp. 95-100

Dinman, B.D., Bovard, W.J., Bonney, T.B., Cohen, J.M. & Colwell, M.O. (1976a) Absorption and excretion of fluoride immediately after exposure - Pt. 1. *J. occup. Med.*, *18*, 7-13

Dinman, B.D., Bovard, W.J., Bonney, T.B., Cohen, J.M. & Colwell, M.O. (1976b) Excretion of fluorides during a seven-day workweek - Pt. 2. *J. occup. Med.*, *18*, 14-16

Dinman, B.D., Backenstose, D.L., Carter, R.P., Bonney, T.B., Cohen, J.M. & Colwell, M.O. (1976c) A five-year study of fluoride absorption and excretion - Pt. 3. *J. occup. Med.*, *18*, 17-20

Dinman, B.D., Elder, M.J., Bonney, T.B., Bovard, P.G. & Colwell, M.O. (1976d) A 15-year retrospective study of fluoride excretion and bony radiopacity among aluminum smelter workers - Pt. 4. *J. occup. Med.*, *18*, 21-23

Discher, D.P. & Breitenstein, B.D., Jr (1976) Prevalence of chronic pulmonary disease in aluminum potroom workers. *J. occup. Med.*, *18*, 379-386

Evang, K. (1938) Investigation among Norwegian workman as to the occurrence of bronchial asthma, acute cryolite poisoning and fluorosis (Norw.). *Nord. Hyg. Tidsskr.*, *19*, 117-148

Field, G.B. & Milne, J. (1975) Occupational asthma in aluminium smelters. *Aust. N.Z. J. Med.*, *5*, 475

Frostad, A.W. (1936) Fluorine intoxication in Norwegian aluminium plant workers (Norw.). *Tidsskr. Nor. Laegefor.*, *56*, 179-182

Gibbs, G.W. (1983) *Mortality experience in Eastern Canada*. In: Hughes, J.P., ed., *Health Protection in Primary Aluminium Production*, Vol. 2, *Proceedings of a Seminar, Montreal, 22-24 September, 1981*, London, International Primary Aluminium Institute, pp. 59-69

Gibbs, G.W. & Horowitz, I. (1979) Lung cancer mortality in aluminium reduction plant workers. *J. occup. Med.*, *21*, 347-353

Giovanazzi, A. & D'Andrea, F. (1981) Mortality of workers in a primary aluminium plant (Ital.). *Med. Lav.*, *4*, 277-282

Gylseth, B. & Jahr, J. (1975) Some hygienic aspects of working in aluminium reduction potrooms with special reference to the use of alumina from the dry-cleaning process of Soderberg pot-gases. *Staub-Reinhalt. Luft*, *35*, 430-432

Haugen, A.A., Becher, G. & Bjørseth, A. (1983) Biological monitoring of occupational exposure to polycyclic aromatic hydrocarbons (PAHs) in an aluminium plant (Abstract). *Eur. J. Cancer clin. Oncol.*, *19*, 1287

Hodge, H.C. & Smith, F.A. (1977) Occupational fluoride exposure. *J. occup. Med.*, *19*, 12-39

Hughes, J.P., ed. (1977) *Health Protection in Primary Aluminium Production. Proceedings of a Seminar, Copenhagen, 28-30 June 1977*, London, International Primary Aluminium Institute

IARC (1982) *IARC Monographs on the Evaluation of the Carcinogenic Risk of Chemicals to Humans*, Vol. 27, *Some Aromatic Amines, Anthraquinones and Nitroso Compounds, and Inorganic Fluorides Used in Drinking-water and Dental Preparations*, Lyon, pp. 255-257

IARC (1983) *IARC Monographs on the Evaluation of the Carcinogenic Risk of Chemicals to Humans*, Vol. 32, *Polynuclear Aromatic Compounds, Part 1, Chemical, Environmental and Experimental Data*, Lyon

International Labour Office (1980) *Occupational Exposure Limits for Airborne Toxic Substances*, 2nd rev. ed. (*Occupational Safety & Health Series 37*), Geneva, pp. 118-119

International Primary Aluminium Institute (1982) *The Measurement of Employee Exposures in Aluminium Reduction Plants*, London,

Jahr, J., Norseth, T., Rodahl, K. & Wannag, A. (1974) *Fluoride exposure of workers in different types of aluminum smelters*. In: *Light Metals. Proceedings of the 103rd AIME Annual Meeting 1974*, New York, American Institute of Mining, Metallurgical & Petroleum Engineers (AIME), pp. 209-236

Johnson, W.M., Shuler, P.J., Curtis, R.A., Wallingford, K.M., Mangin, H.J., Parnes, W. & Donaldson, H.M. (1973) *Industrial Hygiene Survey, Ormet Corporation Aluminum Facilities, Hannibal, Ohio*, Cincinnati, OH, National Institute for Occupational Safety & Health

Kaltreider, N.L., Elder, M.J., Cralley, L.V. & Colwell, M.O. (1972) Health survey of aluminum workers with special reference to fluoride exposure. *J. occup. Med.*, *14*, 531-541

Konstantinov, V.G. & Kuz'minykh, A.I. (1971) Tarry substances and 3:4-benzpyrene in the air of electrolytic shops of aluminum works and their carcinogenic significance. *Hyg. Sanit.*, *36*, 368-371

Konstantinov, V.G., Simakhina, P.G., Gotlib, E.V. & Kuz'minykh, A.I. (1974) *Problem of the carcinogenic hazard in aluminium electrolysis halls*. In: *Occupation Cancers*, Moscow, Academy of Medical Sciences, pp. 87-91

Kreyberg, L. (1959) 3:4-Benzpyrene in industrial air pollution: Some reflexions. *Br. J. Cancer*, *13*, 618-622

Krøkje, A., Tiltnes, A., Mylius, E. & Gullvåg, B. (1984a) Testing for mutagens in an aluminium plant. I. Filter-samples from the working atmosphere. *Scand. J. Work Environ. Health* (in press)

Krøkje, A., Tiltnes, A., Mylius, E. & Gullvåg, B. (1984b) Testing for mutagens in an aluminium plant. II. Expectorates from exposed workers. *Scand. J. Work Environ. Health* (in press)

Lindstedt, G. & Sollenberg, J. (1982) Polycyclic aromatic hydrocarbons in the occupational environment, with special reference to benzo[*a*]pyrene measurements in Swedish industry. *Scand. J. Work Environ. Health*, *8*, 1-19

Machle, W., Thamann, F., Kitzmiller, K. & Cholak, J. (1934) The effects of the inhalation of hydrogen fluoride. I. The response following exposure to high concentrations. *J. ind. Hyg.*, *16*, 129-145

Maestrelli, P., Marcer, G. & Clonfero, E. (1981) Report of five cases of 'fluoride asthma' (Ital.). *Med. Lav.*, *4*, 306-312

Mayer, L. & Guelich, J. (1963) Hydrogen fluoride (HF) inhalation and burns. *Arch. environ. Health*, *7*, 445-447

Midttun, O. (1960) Bronchial asthma in the aluminium industry. *Acta allergol.*, *15*, 208-221

Milham, S., Jr (1976) Cancer mortality pattern associated with exposure to metals. *Ann. N.Y. Acad. Sci.*, *271*, 243-249

Milham, S., Jr (1979) Mortality in aluminum reduction plant workers. *J. occup. Med.*, *21*, 475-480

Møller, P.F. & Gudjonsson, S.V. (1932) Massive fluorosis of bones and ligaments. *Acta radiol.*, *13*, 269-294

Mylius, E.A. & Gullvåg, B. (1984) Alveolar macrophage count (AM-test) as indicator of lung reaction to air pollution. *Acta cytol.* (in press)

Pearson, T.G. (1955) *The Chemical Background of the Aluminium Industry* (*Monograph No. 3*), London, The Royal Institute of Chemistry

Ravier, E.F. (1977) *Technology of alumina reduction*. In: Hughes, J.P., ed., *Health Protection in Primary Aluminium Production, Proceedings of a Seminar, Copenhagen, 28-30 June 1977*, London, International Primary Aluminium Institute

Ricca, P.M. (1966) Exposure criteria for fluorine rocket propellants. Occupational and non-occupational considerations. *Arch. environ. Health*, *12*, 399-407

Ricca, P.M. (1970) A survey of the acute toxicity of elemental fluorine. *Am. ind. Hyg. Assoc. J.*, *31*, 22-29

Rockette, H.E. & Arena, V.C. (1983) Mortality studies of aluminum reduction plant workers: Potroom and carbon department. *J. occup. Med.*, *25*, 549-557

Roholm, K. (1937) *Fluorine Intoxication. A Clinical-Hygienic Study, with a Review of the Literature and Some Experimental Investigations*, London, H.K. Lewis, pp. 121-210

Šaric, M., Gomzi, M., Hruslic, O., Paukovic, R. & Rudan, P. (1979) Respiratory impairment in the electrolytic extraction of aluminium. *Int. Arch. occup. environ. Health*, *42*, 217-221

Sheehy, J.W. (1983) *Occupational Health Control Technology for the Primary Aluminium Industry* (*DHHS (NIOSH) Publ. No. 83-115*), Cincinnati, OH, National Institute for Occupational Safety & Health

Shuler, P.J. & Bierbaum, P.J. (1974) *Environmental Surveys of Aluminum Reduction Plants* (*HEW (NIOSH) Publ. No. 74-101*), Cincinnati, OH, National Institute for Occupational Safety & Health

Simonato, L. (1981) Carcinogenic risk in the aluminium production industry: An epidemiological overview. *Med. Lav.*, *4*, 266-276

Sjögren, B., Lundberg, I. & Lidums, V. (1983) Aluminium in the blood and urine of industrially exposed workers. *Br. J. ind. Med.*, *40*, 301-304

Smith, M.M. (1977) *The respiratory condition of potroom workers: Australian experience*. In: Hughes, J.P., ed., *Health Protection in Primary Aluminium Production*, London, International Primary Aluminium Institute, pp. 79-85

Steinegger, A.F. (1981) Exposure to polycyclic aromatic hydrocarbons in an anode plant and potrooms. *Med. Lav.*, *4*, 259-265

Taylor, W. (1978) Manufacturing processes. Aluminium manufacture. *J. Soc. occup. med.*, *28*, 25-26

Tedder, Y.R. & Chaschin, V.P. (1975) On the effect of alumina used in the electric metallurgy of aluminium on the air of electrolytic departments (Russ.). *Gig. Tr. prof. Zabol.*, *4*, 5-8

Thériault, G., Cordier, S. & Harvey, R. (1980) Skin telangiectases in workers at an aluminum plant. *New Engl. J. Med.*, *303*, 1278-1281

Thériault, G., De Guire, L. & Cordier, S. (1981) Reducing aluminum: An occupation possibly associated with bladder cancer. *Can. med. Assoc. J.*, *124*, 419-423

Vale, J.R. (1981) Occurrence and detection of respiratory disorders in the primary aluminium industry. *Med. Lav.*, *4*, 295-300

van Voorhout, H.C. (1977) *The respiratory condition of potroom workers: Netherlands experience.* In: Hughes, J.P., ed., *Health Protection in Primary Aluminium Production*, London, International Primary Aluminium Institute, pp. 91-93

Wigle, D.T. (1977) Bladder cancer: Possible new high-risk occupation. *Lancet*, *ii*, 83-84

COAL GASIFICATION

1. Historical Perspectives

This subject has been reviewed by Vorres (1979), Huebler and Janka (1980) and Williams (1981).

Coal has been mined and used as a fuel since the beginning of history. When and by whom it was discovered that a combustible gas could be produced if coal was heated in a closed retort is not recorded.

In 1709, Abraham Darby discovered a way of smelting iron ore with coke (Perch, 1979). This was the birth of modern iron and steel production. The industry developed with the growth of the coking industry. The possibilities of coal gas as a source of illumination superior to oil and candles were soon realized, but large-scale use had to wait until the next century. The initiative came from France, where Philippe Lebon had been studying the generation of gas by the destructive distillation of wood and other combustible material. In 1801, he took out a patent for the method.

In 1802, a foundry in Birmingham, UK, was lit by two gas flares and, later, the firm Boulton & Watt obtained a contract to light a cotton mill. The gas was made in six cast-iron retorts and stored in iron, water-sealed holders. The installation was of importance as it also demonstrated the practicability of transmitting gas for district use and was the precursor of the gas industry. In 1812, the Gas Light & Coke Company came formally into existence to supply the cities of London and Westminster and the London Borough of Southwark. By the middle of the nineteenth century, there were few large towns and cities in Europe and North America that did not have gas-works and a system for distributing gas. For a time, factories and large institutions continued to make their own gas but distribution systems soon grew, town became linked with town, and production stations grew bigger, with pipelines transmitting gas over wide areas, in large quantities and at high pressures.

A variation of coal-derived gas appeared in the late 1800s when water-gas generators were introduced. Air and steam were blown over hot coke in a cyclic manner, to produce carbon monoxide and hydrogen. Although of low calorific value, this mixture, water gas, was used for heating or power generation. In a refinement of this process, the waste heat from the air-blow phase could be used to crack gas oil thermally. The resulting decomposition products increased the calorific value of the gas. The gas was called 'carburetted water gas'.

Producer-gas reactors, which appeared at about the same time, also converted essentially all the coke to gas. Producer gas contained about two-thirds nitrogen (from the air) and one-third carbon monoxide (Williams, 1981).

Until about 1960, large amounts of coal were carbonized for the production of town gas in countries where natural gas was not available.

The gas- and coke-production industries progressed independently. The latter specialized in producing metallurgical coke; the coal gas was a by-product and was used to heat the oven.

It was not until the twentieth century that surplus coke-oven gas was used to augment town gas supplies.

A refined process of coal gasification used oxygen instead of air and was operated under pressure. The first full-scale Lurgi gasifier was introduced in 1936 (Parekh, 1982). The Lurgi gasifiers produced a gas of medium-calorific value, which could be used for town-gas production or as a synthesis gas in the chemical industry. After catalytic methanation the product had a composition similar to natural gas. Currently, there are in the order of 100 Lurgi gasifiers (Huebler & Janka, 1980) and other new gasification processes in commercial operation.

In the older gas-making processes, gas and tars were produced by the destructive distillation of coal in a hypoxic environment. In the more modern gasification processes, air or oxygen is introduced in sub-stoichiometric amounts resulting in partial combustion of the coal. This provides heat, which breaks down the rest of the coal, which then reacts with steam to form gas. Coke is not produced as a by-product by the modern gasifiers.

Estimate of the current number of gasification workers

Coal is a major world commodity. In 1976, coke production amounted to more than 367 million tonnes (Perch, 1979), and the current world output of crude coal-tars is estimated at about 18 million tonnes (McNeil, 1983). In the USA, about 15% of the mined coal is used for gasification and coke production (Vorres, 1979). Some coke-oven gas is also consumed as town gas. The production rate of gas-works is difficult to estimate because the number of small units is large and most plants are rapidly switching to natural gas or oil.

In 1936, about 3800 mechanical and 1000 hand-fed producers were installed in the USA; about 2000 mechanical and 600 hand-fed machines were in continuous or periodic use (van der Hoeven, 1945).

The Electric Power Research Institute (EPRI) (1983) has attempted to prepare a partial list of modern coal-gasification plants. Their listed plants use a total of about 60 000 tonnes of coal per day. About 2000 workers in the USA have been engaged in pilot-plant gasifiers in the recent past.

The Working Group considered both the older gas-works and the newer coal-gasification systems. While some epidemiological data exist for exposures associated with the older retorts producing town gas, detailed chemical characterization and short-term toxicological test data are available only for the newer systems.

2. Description of the Processes

2.1 Town-gas retorts

(a) Coal distillation

The original gasifiers were retorts of two types - horizontal and vertical. The horizontal was the older of the two and operated on a batch basis. For town-gas production, typical horizontal retorts were refractory structures about 6 m long and shaped like an inverted 'U', with a flat base 0.6 m wide by 0.4 m high. They were closed at both ends with a cast-iron door and set horizontally in groups of 8 or 10 on either side of a combustion chamber in which hot producer gas was burnt. Each retort held about two-thirds of a tonne of coal, which was heated to 1000-1200°C for 8-10 hours. The evolved gases and vapours were drawn off. On completion, the residual coke was forced out with a ram and the retort was recharged. If the demand for gas was low, the retorts could be left uncharged. During charging and discharging, dusts and fumes were evolved. Retorts were later built vertically, which allowed the discharge to take place by gravity and simplified the charging machinery.

A more efficient process for town-gas and coke production resulted from the introduction of continuous vertical retorts (Williams, 1981). These retorts are about 8 m high and rectangular in shape, about 2.5 m by 0.25 m at the top and slightly wider at the bottom, and they are usually arranged in rows. Coal is fed into the retort by a valve or lock from hoppers over the top, and coke is extracted from the bottom. The process is continuous, and manual labour is needed only occasionally when charges stick and have to be removed by rodding. Steam can be injected into the retort to improve the thermal efficiency, and oil can also be injected to raise the calorific value.

(*b*) *Gas purification*

The gas produced by carbonization in a retort was a brown smoky vapour known as 'foul gas'. The first step in its purification was to condense out the higher boiling fractions in condensers that were often little more than towers in which the gas flow was slowed and cooled and brought into contact with baffle plates. Later, more efficient electrostatic precipitators were used. The next step was to wash the gas with water to remove the ammonia. Finally, sulphur, present mainly as hydrogen sulphide, had to be removed. Originally, this was done with lime, resulting in calcium sulphide piles with an offensive smell. This was superceded by the use of a granular type of iron oxide known as 'bog oxide', which could be partially regenerated by exposure to air and which, when the oxide was finally spent, could be used to make sulphuric acid.

Although the coal gas could be used directly, it was usually mixed with other gases, because of the necessity of maintaining calorific value and burning characteristics within a fairly narrow range. The most common additive was 'water gas', a mixture of carbon monoxide and hydrogen made by passing steam over red-hot coke. This process was endothermic and cyclic. When the coke began to cool, the steam was shut off and air was admitted to the coke. As the coke burned, the temperature was raised until it was sufficiently hot to allow steam to be passed over it again. In a refinement of the process, the hot gases from the burning coke were used to heat an open brickwork onto which gas oil (a petroleum oil) was sprayed to be thermally cracked, making carburetted water gas. Producer gas, a mixture of hydrogen, carbon monoxide, carbon dioxide and nitrogen gases, made by burning coke in a restricted air supply with an injection of steam, was occasionally used as an additive but owing to its low calorific value tended to dilute rather than enrich the gas. Mixtures of butane or propane and air were often added and, when available, suitable stack gases from refineries and chemical works were also used. The resulting gas was distributed as town gas.

2.2 Modern coal-gasification processes

(*a*) *Introduction*

This subject was reviewed by Lowry (1945, 1963) and by Braunstein *et al.* (1977).

Coals are ranked by the amount of volatile matter obtained upon heating and range from anthracites, through bituminous to lignite coals. The amount of volatile matter in anthracites varies from 3-14% by weight on a dry, ash-free basis; that in bituminous coals varies from 14% to over 30%; lower-rank coals are classified by their calorific value. The carbon content of coals varies from over 90% by mass for anthracite to about 70% for lignite coals.

All coals are contaminated with minerals, trace metals and compounds containing sulphur, nitrogen and oxygen. Sulphur is commonly found in pyrite inclusions in coal, as well as in thiophenic-like structures in the organic portion of the coal itself. Nitrogen can be present in pyridine, quinoline, amine and pyrrole-like organic structures in the coal. Most coal will contain 5-20% mineral matter by weight. This includes layered alumino-silicates, calcium, iron and magnesium carbonates, silica, pyrite, pyrrhotite and other iron sulphides, and sulphates. These contaminants are also found in the by-products from the gasification processes (described below).

Coal consists mainly of carbon and hydrogen and can be regarded as a highly complex polynuclear compound with sulphur-, nitrogen- and oxygen-containing radicals. It is deficient in hydrogen and, for total gasification, steam is necessary to provide the hydrogen to make up this deficiency.

In modern coal gasification, air or oxygen and steam are reacted with coal under various conditions of temperature, flow and residence time to oxidize partially the coal and produce combustible gases, in which the major fuel constituents are carbon monoxide, hydrogen and methane. The coal is completely consumed in modern processes, i.e., no coke is produced. The major chemical reactions that occur in a coal gasifier are as follows (Ondich, 1983):

$$C + H_2O + heat \rightarrow CO + H_2$$
$$C + CO_2 + heat \rightarrow 2CO \qquad \text{(gasification)}$$
$$C + 2H_2 \rightarrow CH_4 + heat$$

$$2C + O_2 \rightarrow 2CO + heat$$
$$C + O_2 \rightarrow CO_2 + heat \qquad \text{(combustion)}$$

When an air-steam mixture is used to gasify the coal, the product gas is of low calorific value, containing nitrogen as a major component. This gas is suitable for use as a fuel near its point of generation but it is not economical for long-distance transmission. Gas of medium calorific value (which contains a minor amount of nitrogen) is obtained when oxygen and steam are used to gasify the coal. This gas can be used either as an energy source or as a synthesis gas for the production of chemicals or liquid fuels.

Gas of medium calorific value can be further processed to produce a synthetic pipeline gas of high calorific value, which contains over 90% methane and is very similar to natural gas. Production of methane requires a ratio of hydrogen to carbon monoxide of 3:1. To obtain this ratio, the gases are shifted by reaction carbon monoxide and water:

$$CO + H_2O \rightarrow CO_2 + H_2 \qquad \text{(water-gas shift reaction)}$$

and, in the presence of catalysts, hydrogen is then reacted with carbon monoxide to form methane:

$$3H_2 + CO \rightarrow CH_4 + H_2O.$$

The first stage in total gasification is the destruction of the molecular structure of the coal by heat: the coal swells and becomes plastic, and any volatile compounds formed evaporate.

According to the conditions in the chamber, a proportion of these are swept out unchanged as tar and oils with the process gas. The remainder, which is mainly carbon, is partially oxidized, providing most of the heat by which the reaction continues. The reactions taking place within the chamber are extremely complex, and those shown above represent ideal reactions leading to the desired state. Higher temperatures in the chamber favour these ideal reactions, but practical considerations limit the temperatures that can be used.

The by-products generated in various coal-gasification processes are of three types: gases such as hydrogen sulphide, carbonyl sulphide, ammonia and hydrogen cyanide; hydrocarbon gases, aerosols and residues, including polynuclear aromatic compounds; and mineral particulates and wastes. Most of the sulphur and nitrogen in the coal are released as reduced gases, principally hydrogen sulphide and ammonia. Coal-gasification products are usually subjected to one of a number of processes to remove the hydrogen sulphide.

The production of tars and gases during gasification is, to a great extent, determined by the conditions and dynamics of the gasification process. In general, high-temperature, entrained-bed, co-current processes (in which coal and gas flow are in the same direction) produce little heavy hydrocarbon (tar) by-products, while fixed-bed, counter-current processes (in which coal and gas flow are in opposite directions) may produce tar by-products.

When coal is fed counter-current to gas flow, as in a fixed-bed gasifier, the preheat zone or the region at the top of the coal bed evolves hydrocarbon gases, oils and tars through the release and decomposition of volatile organic matter from the coal. The persistence or decomposition of heavier hydrocarbons is dependent on a number of factors, including type of coal, gasifier design and operating temperature, pressure and residence time of the tar in various gasifier regions (van der Hoeven, 1945; von Fredersdorff & Elliott, 1963). Fixed-bed gasifiers may produce 4 to 40 litres of tar per tonne of coal gasified (von Fredersdorff & Elliott, 1963).

(*b*) *Commercial gasifiers* (Parekh, 1982)

For the most part, modern coal gasifiers may be assigned to three general types of process: fixed-bed, fluidized-bed and entrained-bed (Huffstetler & Rickert, 1977). Fixed-bed gasification typically produces much greater quantities of tars than fluidized-bed or entrained-bed gasificaton.

Other factors that have a major influence on the gasification process are pressure (1-30 atmospheres), oxidant (air or oxygen), and design and operational control of process temperatures, which (in the case of fixed-bed and fluidized-bed processes) determine whether or not released mineral matter is slagged (melted).

A diversity of downstream processing operations may be used with the gasifier. Their selection is based on the characteristics of the gas leaving the gasifier, environmental constraints, and the quality of end product required. The gas has been used or may be used for: industrial and town gas; synthesis of gas of high-calorific value (synthetic natural gas); indirect liquefaction (Fischer-Tropsch liquefaction) (Huffstetler & Rickert, 1977); methanol synthesis; ammonia production; or combined-cycle electrical power generation (Electric Power Research Institute, 1983).

A few examples of gasification systems were selected for review by the Working Group on the basis of their commercial use or their use in subsequently described studies of process characterization or industrial hygiene. A complete Lurgi gasification system is described below.

However, it is emphasized that there is wide variation in the general design and the specifications of the unit operations both within and among different gasification systems. Selection of a system depends on a number of parameters, including the type and end use of the gas produced; e.g., methane, methanol, combined-cycle power generation.

(i) *Fixed-bed gasifier - Lurgi* (National Institute for Occupational Safety & Health, 1978; Young *et al.*, 1978; Young & Evans, 1979)

The Lurgi gasifier (shown in Fig. 1) is a fixed-bed gasifier, pressurized (to about 30 atmospheres) and fed with coal (screened to a lump size of 0.5-5.0 cm), oxygen and steam. Since 1936, approximately 100 such gasifiers have been deployed.

Fig. 1. Lurgi gasifier (from Huffstetler & Rickert, 1977)

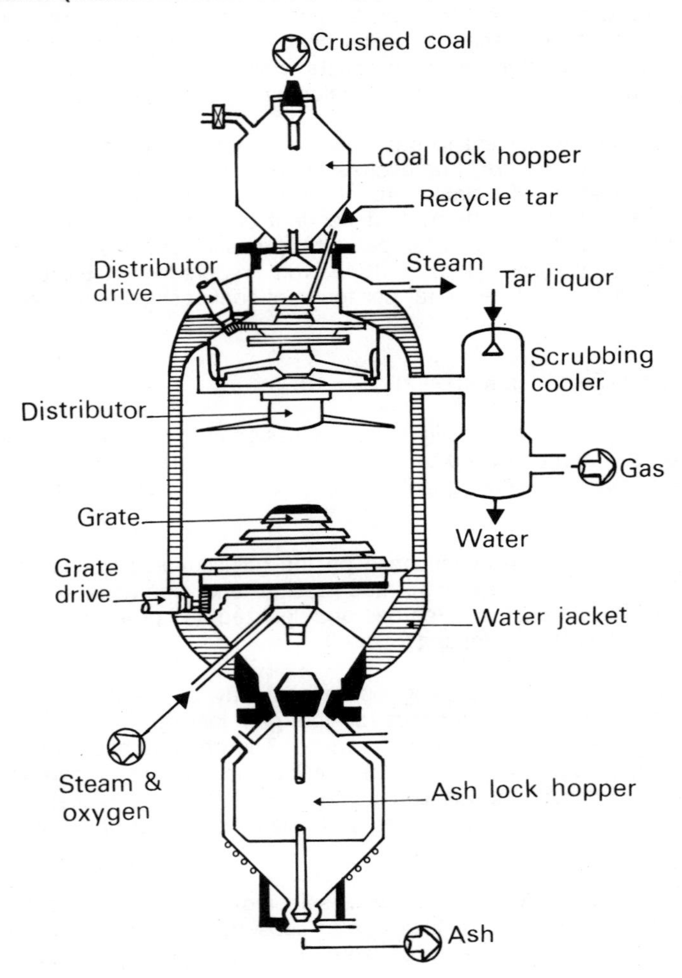

A complete Lurgi coal-gasification system for the production of synthetic pipeline gas comprises the following operations: coal handling and preparation, coal feeding, coal gasification, ash removal, gas quenching and cooling, shift conversion, gas purification, methanation, sulphur removal, gas-liquor separation, phenol and ammonia recovery, and by-product storage and cleanup (shift conversion and methanation are needed only for the final processing to produce synthetic pipeline gas). A generalized scheme for the production of a gas of high-calorific value is presented in Figure 2, and the operations are described below. Some of the steps that might result in hazardous exposures owing to fugitive (leaking) emissions or to the nature of the handling processes are reviewed here. Five operations: coal preparation, injection, gasification, quenching and methanation are of particular concern for health (Young & Evans, 1979).

Fig. 2. Coal gasification operations for the production of synthetic pipeline gas (from Young *et al.*, 1978)

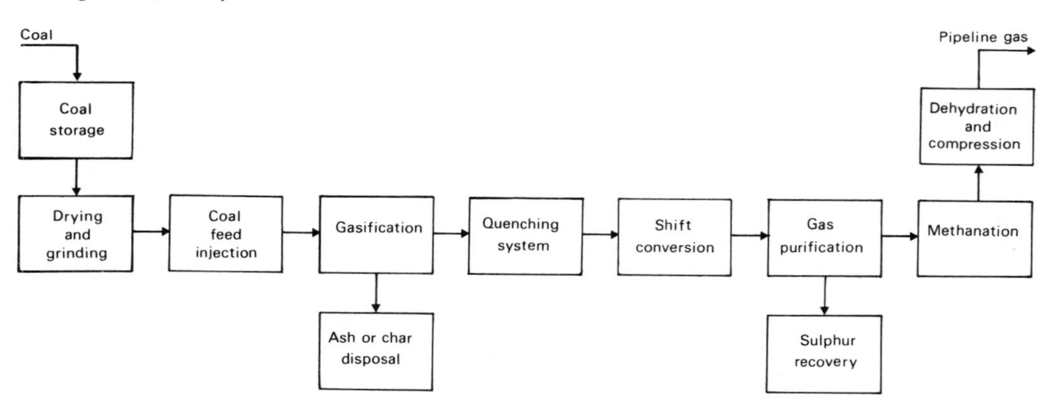

Coal handling and preparation: Coal is delivered from the mine to the plant's unloading hopper, from which it is transferred by feeders and conveyors to crushers. The preparatory steps also include cleaning, drying, pulverizing and stockpiling operations.

Coal feeding: After preparation, the coal is fed from the gasifier's coal bunker to the lockhopper by gravity. If the gasification is performed under pressure, the lockhopper is first charged with coal, pressurized, then opened to discharge the coal into the gasifier, closed, depressurized and recharged. It is conceivable that raw gas could be released into the work environment through leakage or malfunctioning equipment at this point.

Gasifier operation: Lurgi operation is typical of fixed-bed gasifier operations in that coal is fed to the top of a slowly descending bed, while oxidant gas and steam are introduced into the bottom of the gasifier. At the top of the gasifier coal bed, coal is distilled by rising hot gases from below. This releases tars and hot oils into the product gas stream in amounts of the order of 45 kg for each tonne of (bituminous) coal consumed (Parekh, 1982). Beneath the zone of pyrolysis in the bed, the carbon in the residue is gasified at 650-815°C by the hot gases and steam rising from below. At the bottom of the bed, the remaining carbon is oxidized at 980-1370°C, providing the hot gases for the reactions in the higher zones.

Ash removal: Ash (mineral residue from the coal) from the gasifier is continuously removed by a rotating grate and collected in a pressurized lockhopper from which it is discharged. At the end of the ash-discharge cycle, the lockhopper is depressurized. The mineral composition of typical coal ash, in this case from a Lurgi gasifier, is given in Table 1.

Table 1. Concentrations of trace elements in Illinois coal and in the unquenched gasifier ash obtained from it after Lurgi gasification[a]

Element	Concentration (mg/kg)	
	In coal (Illinois No. 5)	In ash
Antimony	0.1	19
Arsenic	1.9	0.3
Barium	-	490
Beryllium	2.0	19.8
Boron	307	673
Bromine	6.6	-
Cadmium	<0.3	<0.3
Cerium	-	41
Cesium	-	11
Chromium	15	592
Cobalt	3.8	38
Copper	10	273
Fluorine	55	4.6
Iron	1.3[b]	15[b]
Lanthanum	3.6	42
Lead	28.1	200
Lithium	5.5	54
Manganese	23	305
Mercury	0.17	0.016
Molybdenum	7	8
Nickel	32	462
Potassium	1.3[c]	14[c]
Scandium	1.6	-
Selenium	9	-
Silver	0.3	3.0
Sodium	2.8[d]	29[d]
Tantalum	-	1.3
Vanadium	21	181
Ytterbium	-	11
Zinc	2.4[d]	16[d]

[a]From Braunstein *et al.* (1977)
[b]Concentration x 10^4
[c]Concentration x 10^3
[d]Concentration x 10^2

Gas quenching and cooling: Raw gas from the gasifier is quenched with recycled gas liquor (described below) in a wash cooler. The quenched gas then undergoes primary cooling in waste-heat boilers and is further cooled in a series of indirect coolers. Moisture, tars, oils and phenols are condensed as the raw gas passes through the wash cooler and primary and secondary coolers. The resulting condensate, termed gas liquor, is sent to the gas-liquor separation unit. Also present are water-soluble components from the raw gases, including ammonia and hydrogen cyanide, and dust scrubbed from the raw gases.

Gas-liquor separation: The gas-liquor separation unit consists of a series of coolers, depressurization vessels and gravity separators. At low pressure, the gases absorbed in liquids come out of solution. Tars and oils are separated from the aqueous phase of the gas liquor and recovered as by-products. The tar-oil component is either fractionated for sale as a refined product or burned for its calorific value. The tars may either be returned to the gasifier for further cracking or used for coal-tar production. The gas-liquor fraction is processed to recover phenol and ammonia.

Phenol and ammonia recovery (Dailey, 1977): Phenol recovery utilizes an organic solvent such as isopropyl ether or *n*-butylacetate to remove phenolic compounds from the gas-liquor stream. Phenols are recovered from the solvent by steam stripping and are either stored or further refined for subsequent sale. The phenol-free gas liquor is steam stripped to remove acid gases. The same liquor is then steam stripped at a higher temperature to remove ammonia, which is condensed for sale.

Gas purification: Ammonia has usually been removed from the gas stream by water quenching and scrubbing before the gas reaches this stage. A number of purification processes for hydrogen sulphide have been developed and deployed extensively in commercial operations, using organic solvents, amines, glycols, aqueous salt solutions or fixed-bed absorbants. The removed hydrogen sulphide is processed further to recover the sulphur as dilute sulphuric acid or elemental sulphur.

Hydrogen sulphide disposal: The conventional technique for disposal of the hydrogen sulphide stream from the purification unit is the Claus process. One-third of the hydrogen sulphide gas is burned to sulphur dioxide and recombined, in the presence of bauxite, with hydrogen sulphide to form sulphur and water. From the reactor, the hot gases flow into the condenser, where they are cooled and the liquid sulphur is removed.

Water-gas shift conversion: Prior to methanation (described below), the ratio of hydrogen to carbon monoxide must be adjusted to approximately 3:1. This is done by passing a portion of the gas through the shift-conversion unit, where carbon monoxide reacts with steam at 400°C to form carbon dioxide and hydrogen in the presence of commercial catalysts, which contain either iron-chromium oxides or cobalt-molybdenum oxides on an aluminium base. In addition, desulphurization and hydrogenation of organic compounds may occur.

Methanation: Methane synthesis is accomplished with a nickel catalyst, by the reaction of carbon monoxide with hydrogen to produce methane and water. The catalyst contains 25-77% nickel on a porous refractory support, such as alumina or kieselguhr and it has to be replaced after two to three years of use.

(ii) *Fluidized-bed gasifier - Winkler* (Huffstetler & Rickert, 1977; Parekh, 1982)

This fluidized-bed gasifier has seen wide deployment in past decades for the production of synthesis gas and ammonia. Since its development in 1926, some 60 gasifiers have been installed, of which about 30 were still in service in 1980.

Air or oxygen is blown up from the bottom of the coal bed at a velocity great enough to entrain or 'fluidize' partially the bed of coal. Feed coal is crushed to permit fluidization at reasonable gas velocities. The thorough mixing results in bed temperatures of near 1090°C and in the gasification of most of the tars distilled from the coal. It produces a gas of low calorific value.

(iii) *Entrained-bed gasifier - Koppers-Totzek* (Huffstetler & Rickert, 1977; Huebler & Janka, 1980)

Since 1950, over 50 Koppers-Totzek gasification plants have been constructed or are under construction in Europe, Africa, Asia and South America (Huebler & Janka, 1980).

Pulverized coal is reacted with co-currently flowing oxygen, and a temperature of about 1900°C is reached. Because of the fine size of the coal particles, as well as the co-current, high-temperature flow, essentially complete gasification of the coal-distilled tars is possible.

(iv) *Underground coal gasification* (Huffstetler & Rickert, 1977)

Underground gasification is used on an industrial scale in the USSR, and has been the subject of research and development in the USA.

Coal may be gasified underground by drilling to the coal seam and providing sufficient air for partial combustion of the coal, thereby producing heat for endothermic gasification reactions. Products of gasification are drawn off through separate bore-holes. The basic chemical processes involved are the same as for conventional coal gasification.

Different designs may be used to inject air or oxygen and to extract the gasification product. The coal in the seam may be reduced to a rubble, followed by injection and extraction of gases by separate bore-holes into the rubble pile. Alternatively, the seam may be left intact with bore-holes drilled into it some distance apart. With some level of permeability of the seam to gas, the fire and gasification zone may be established at either the air inlet or at the product-gas outlet. In the former case, until large channels burn through, the intact coal seam may act, to some extent, as a filter and absorber for particulates and condensable materials in the product gas. In the latter case, those materials may be carried to the product-gas outlet, so that the product may, in general, resemble the gas produced from fixed-bed gasification.

(*c*) *Experimental gasifiers*

(i) *Fixed-bed gasifier - METC-stirred bed*

The US Department of Energy-Morgantown Energy Technology Center (METC) operates a pilot-scale, fixed bed, pressurized, dry bottom-ash gasifier utilizing a modified Lurgi process (with a coal feed of approximately 20 tonnes/day). Its major features are a stirring arm, to permit the use of any caking coal as fuel, and a versatile, gas clean-up system which is used to test various hot or wet clean-up methods. This pilot plant has been used to characterize tar aerosols produced during gasification and for toxicology studies (Lamey *et al.*, 1981, Royer *et al.*, 1983).

(ii) *Fixed-bed gasifier - Stoic*

The Stoic gasifier is a Foster-Wheeler, two-stage, fixed bed, air-blown gasifier. One used for heating at the University of Minnesota, USA, has been used also for chemical and toxicological characterizations; it provides a gas of low calorific value (Griest & Clark, 1982).

(iii) *Fluidized-bed hydro-gasifier - HYGAS*

The HYGAS process is a three-stage, fluidized bed hydro-gasifier, producing a gas of high-calorific value on a pilot scale, which is used by the US Institute of Gas Technology and the Argonne National Laboratory for technology development and testing. In the top stage of the gasifier, hydrogasification is carried out using the hot fuel gas generated in the low stages. This top-stage, low-temperature reactor produces significant amounts of heavy hydrocarbon compounds compared to single-stage, fluidized-bed gasification using air or oxygen. These materials are condensed out for use in a recycled oil, which is used to prepare the coal as a slurry for feeding to the hydrogasifier. The chemical composition and toxicology of these tars have been studied (Stamoudis *et al.*, 1981).

3. Exposures in the Workplace

Some of the materials to which workers in the coal-gasification industry may be exposed are listed in Table 2.

Table 2. Airborne substances (and classes of substances) found in the coal gasification industry[a]

Material	Principal uses or sources of emission
Common airborne contaminants	
Alcohols	Gasifier operations, gas purification, scrubber solvents
Aldehydes, aliphatic and aromatic	Coke-oven and gasifier operations
Amines, aliphatic and aromatic*	Gasifier operations, ammonia recovery, gas purification, sulphur removal, scrubber solvents
Ammonia	Gasifier operations, ammonia recovery and storage
Bauxite	Shift conversion, catalysts
Carbon	Coal storage, transport, preparation and feeding, coke quenching and handling, ash and soot removal
Carbon dioxide	Gasifier operations, gas quenching and cooling, shift conversion, gas purification and methanation
Carbon disulphide	Gasifier operations, gas quenching and cooling, shift conversion, gas purification, sulphur removal
Carbon monoxide	Gasifier operations, gas quenching and cooling, shift conversion, gas purification and methanation
Carbonyl sulphide	Gasifier operations, gas quenching and cooling, shift conversion, gas purification, sulphur removal
Carboxylic acids	Gasifier operations, gas quenching and cooling, gas-liquor separation
Hydrocarbons*, aliphatic, cyclic and aromatic	Gasifier operations, gas quenching and cooling, gas-liquor separation, oil and tar storage, shift conversion, gas purification, sulphur removal, gas and by-product handling
Hydrogen cyanide	Gasifier operations, gas quenching and cooling, shift conversion, gas purification
Hydrogen sulphide	Gasifier operations, gas quenching and cooling, shift conversion, gas purification, sulphur removal
Mercaptans	Gasifier operations, gas quenching and cooling, shift conversion, gas purification
Nitrogen heterocyclics	Gasifier operations, gas quenching and cooling, gas-liquor separation, ammonia recovery
Nitrogen oxides	Gasifier operations
Oxygen heterocyclics	Gasifier operations, gas quenching and cooling, gas-liquor separation
Phenols*	Gasifier operations, gas quenching and cooling, gas-liquor separation, phenol recovery, oil and tar storage
Polynuclear aromatic hydrocarbons*	Gasifier operations, gas quenching and cooling, gas-liquor separation, oil and tar storage

Material	Principal uses or sources of emission
Silica	Coal storage, transport, preparation and feeding, ash removal, refractory materials
Sulphur dioxide	Gasifier operations, gas purification, sulphur removal
Sulphur heterocyclics	Gasifier operations, gas quenching and cooling, gas-liquor separation, sulphur removal
Sulphuric acid	Gas purification, ammonia recovery, scrubber chemicals
Thiocyanates	Gasifier operations, gas quenching and cooling, shift conversion, gas purification, sulphur removal
Other contaminants	
Arsenic, oxides and salts*	Gasifier operations, ash removal, gas purification, sulphur removal, scrubber chemicals
Asbestos*	Thermal insulation, clothing
Cadmium and cadmium oxides*	Ash removal
Glycols	Gas purification, sulphur removal, scrubber chemicals
Lead and lead oxides*	Ash removal
Nickel, metal and oxides*	Methanation, catalysts
Nickel carbonyl*	Methanation
Vanadium, oxides and salts	Ash removal, gas purification, sulphur removal, scrubber chemicals

[a]This list includes chemicals or classes of chemicals used in or formed during coal gasification, and the processes during which they are used or formed or during which exposures are most likely to occur. It was compiled from information collected during the preparation of this monograph and cannot pretend to be exhaustive.

*Chemicals (and classes of chemicals) indicated by an asterisk have been considered previously in the *IARC Monographs* series and are listed in the Appendix, together with the evaluation of their carcinogenicity.

(*a*) *Town-gas retorts*

Lawther *et al.* (1965) measured benzo[*a*]pyrene levels in the air in gas-works retort houses. A mean concentration of 4.8 μg/m³ was found in a horizontal-retort house as against 1.4 and 2.7 μg/m³ in two different vertical-retort houses. It was estimated that the concentration of benzo[*a*]pyrene to which men working in retort houses were exposed was 100 times that found in urban air (the City of London). The mean concentration of benzo[*a*]pyrene in a retort house was about 3 μg/m³, but the top man (the man working directly above the retort) in a horizontal-retort house could be exposed to 200 μg/m³. Table 3 presents the concentrations

Table 3. Mean concentrations of suspended matter and polynuclear aromatic hydrocarbons in the retort-house air of gas-works in the UK (long-period samples)[a]

Material	Concentration (μg/m³ of air)		
	Rochester	Brighton (Portslade)	Birmingham (Windsor Street)
Suspended matter[b]	1000	925	1100
Benzo[*e*]pyrene	2.5	0.9	1.4
Coronene	0.5	0.7	0.4
Benzo[*ghi*]perylene	2.5	2.4	1.6
Benzo[*a*]pyrene	4.8	1.4	2.7

[a]From Lawther *et al.* (1965)
[b]Consisting of tars with some coal dust

of suspended matter and polynuclear aromatic hydrocarbons found in air samples from gas-works retort houses. Table 4 presents the amounts of some polynuclear aromatic hydrocarbons extracted from face masks. [The results reported in Table 4 can be used as approximate exposure indices for the various polynuclear aromatic compounds. Other hydrocarbons were detected qualitatively, but the methods of analysis used did not permit more than four of them to be reliably determined, i.e., benzo[*e*]pyrene, coronene, benzo[*ghi*]perylene and benzo[*a*]pyrene.]

Table 4. Mean shift exposure to polynuclear aromatic hydrocarbons in retort houses in the UK (mask samples)[a]

Polynuclear aromatic hydrocarbon	Exposure per eight-hour shift (μg)[b]			
	Rochester	Birmingham (Saltley)	Bristol	Leeds
Benzo[*e*]pyrene	495	27 (23-35)	16 (5-25)	1 (0-6)
Coronene	83	8 (4-18)	4 (0-7)	0
Benzo[*ghi*]perylene	557	33 (29-36)	21 (4-36)	1 (0-5)
Benzo[*a*]pyrene	1136	56 (42-84)	37 (13-61)	3 (0-12)

[a]From Lawther *et al.* (1965)

[b]The numbers of masks analysed at each site were as follows: Rochester, 1; Birmingham (Saltley), 4; Bristol, 3; Leeds, 6. The figures in parentheses refer to the range of concentrations found.

Table 5. The content of polynuclear aromatic compounds in the air (μg/100 m³) in two gas-works in Norway, compared to that in general town air in the UK

Compound	Source of sample				
	Industrial plants[a]		General town air[b]		
	Oslo gas-works	Bergen gas-works	Liverpool	Wrexham	Llangefni
Anthanthrene	80	60	-	-	-
Anthracene	100	Trace	-	-	-
Benz[*a*]anthracene	1400	280	-	-	-
Benzo[*ghi*]perylene	210	200	16.6	5.1	0.5
Benzo[*a*]pyrene	730	200	6.8	2.0	0.4
Benzo[*e*]pyrene	150	130	-	-	-
Coronene	20	30	-	-	-
Fluoranthene	1300	240	6.7	2.0	0.4
Fluorene	20	Trace	-	-	-
Naphthalene	ND[c]	ND	-	-	-
Phenanthrene	ND	8	-	-	-
Pyrene	870	240	5.0	1.8	0.3

[a]From Kreyberg (1959)

[b]From Commins (1958); the measured amount of fluoranthene may include some benz[*a*]anthracene, as the two absorption peaks are very close together

[c]ND, not detected

Battye (1966) sampled retort-house air for amines using a Soxhlet-extractor technique and demonstrated the presence of 1- and 2-naphthylamines. He estimated that they were present at a concentration of 0.002 μg/m³. A sample taken outside the retort house showed a concentration of 0.0003 μg/m³. [The Working Group noted that the two samples were taken and analysed by semi-quantitative methods.]

Kreyberg (1959) determined concentrations of several polynuclear aromatic compounds in two gas-works in Norway, as shown in Table 5. This table additionally contains concentrations found in the general air in three towns in the UK, for comparison (Commins, 1958). It is notable that the concentrations of 4-, 5- and 6-ring compounds were found to be high in relation to the concentrations of 2- and 3-ring compounds. Lindstedt and Sollenberg (1982) reported an average airborne concentration of benzo[*a*]pyrene of 4.3 μg/m³ at the top of the retorts in a town gas-works. This was based on 134 one-hour measurements made in 1964. The maximum concentration measured was 33 μg/m³. At a distance of 25 m from the retorts the concentration was 0.15 μg/m³.

(*b*) *Exposures from modern gasification systems*

Pilot-scale coal gasifiers representative of each of the three generic types of gasification were studied recently (Cubit & Tanita, 1982). The study stressed respirable exposures to toxic gases and to polynuclear aromatic compounds (PACs), excluding amines. Area and personal sampling were performed for PACs, using silver membrane filters and Chromosorb cassettes. Adsorbed and subsequently eluted PACs were analysed using gas chromatography with mass spectrometry with additional analyses by capillary column gas chromatography or high-performance liquid chromatography, as required. Material eluted from the samplers was analysed for 27 PACs.

Total PAC vapour or aerosol concentrations measured in various operation areas for each site are presented in Table 6. PAC concentrations were much greater for the fixed-bed site than for the other two; and the levels were somewhat higher for the fluidized-bed gasifier than for the entrained-bed gasifier. In 1981, no airborne PAC of greater than two-ring structure was found associated with fluidized-bed gasification, and only a small percentage of the PACs at the fixed-bed and entrained-bed locations were of greater than three-ring structure.

Table 6. Total concentrations of polynuclear aromatic compounds for area samples of coal-gasification plants[a]

Plant	Area sampled[b]	N[c]	GM[d] (μg/m³)	GSD[e]
A (entrained-bed)	Gasifier combustor	4	0.4	4.9
	Gasifier reductor	1	0.4	NA[f]
	Gas duct	2	<0.1[g]	<0.1
	Induced-'draft' fan	2	0.3	6.1
	Sludge thickener	1	<0.1	NA
	Cyclone scrubber	1	0.2	NA
	Control room	2	0.2	3.5
	Total in-plant	13	0.2	3.2
	Plant perimeter	2	0.3	4.7

Plant	Area sampled[b]	N[c]	GM[d] (μg/m³)	GSD[e]
B (fixed-bed)	Top of gasifier	3	22.7	1.5
	Poke hole	3	17.3	2.7
	Detarrer/deoiler	4	18.1	2.4
	Oil/liquor separator	1	74.2	NA
	Ash pan	1	42.1	NA
	Day tar/oil tank	2	20.6	2.2
	Tar pump	2	72.4	1.1
	Gas compressor	1	63.6	NA
	Control room	2	20.5	1.8
	Total in-plant	19	26.8	2.2
	Plant perimeter	2	1.8	4.7
C (fluidized-bed)	Gasifier - 5th level	3	0.6	2.6
	Gasifier - 6th level	4	5.4	7.6
	Gas compressor[h]	4	1.7	3.0
	Filter/strainer[h]	2	0.9	2.4
	Control room	4	2.9	3.7
	Total in-plant	17	2.0	4.3
	Plant perimeter	3	1.1	2.6

[a]From Cubit and Tanita (1982)

[b]For precise definitions of the locations see Cubit and Tanita (1982)

[c]N, Number of samples collected

[d]GM, Geometric mean of concentrations

[e]GSD, Geometric standard deviation

[f]NA, not applicable

[g]<, indicates value below detection limit

[h]Samples taken where contamination with coal dust possible

Personal samples from various categories of workers showed that levels associated with fixed-bed gasification were greater than those with the other two (Table 7). The only exception to this was the exposure of a maintenance worker in the fluidized-bed gasification plant. Most exposures were to two- or three-ring PACs.

Table 7. Total concentrations of polynuclear aromatic compounds in personal samples taken at coal-gasification plants[a]

Plant	Personnel sampled	N[b]	GM[c] (μg/m³)	GSD[d]
A (entrained-bed)	Equipment operators	8	2.6	3.6
	Maintenance personnel			
	Welder/boilermaker	4	6.8	1.9
	Pipefitter	2	6.5	3.4
	Labourer	2	0.4	7.7
	Total maintenance	8	3.4	5.1

Plant	Personnel sampled	N[b]	GM[c] (μg/m³)	GSD[d]
	Chemical technician	2	0.4	7.4
	Instrument technician	1	1.9	NA[e]
B (fixed-bed)	Operators	4	19.3	1.5
	Maintenance personnel			
	Millwright	2	58.0	2.5
	Utility helper	2	15.3	1.1
	Maintenance	2	88.5	1.3
	Electrician	1	389.4	NA
	Foreman	1	35.1	NA
	Total maintenance	8	55.1	3.0
C (fluidized-bed)	Lower-level technician	7	4.2	1.3
	Upper-level technician	5	5.3	11.3
	Maintenance personnel	1	60.4	NA

[a]From Cubit and Tanita (l982)

[b]N, Number of samples collected

[c]GM, Geometric mean of concentrations

[d]GSD, Geometric standard deviation

[e]NA, not applicable

Wipe sampling at the three sites qualitatively indicated the potential for dermal exposure to four- and five-ring compounds.

(*c*) *Tars produced by gasifiers*

A number of polynuclear aromatic compounds (PACs) have been identified in tar by-products of coal gasification. Table 8 provides a comparison between tars from a vertical retort, a Lurgi gasifier, a low-temperature gasifier and coke-ovens for several tar fractions or specific PACs (McNeil, 1983).

Quantitative measurements of six polynuclear aromatic hydrocarbons (PAHs) have been made for tar and oil by-products of the commercial-scale Lurgi-gasification facility in Kosovo, Yugoslavia (Ondich, 1983); benzo[*a*]pyrene was found at a concentration of 240 mg/kg in heavy tar, and 68 mg/kg in medium oil.

Over 20 PACs were found in each of the neutral, basic and acidic fractions of gas-stream tar from the METC-stirred bed experimental fixed-bed gasifier, with most being found in the neutral fraction (Lamey *et al.*, 1981). In another study of the same gasifier, a gas-stream tar was separated into 13 fractions or sub-fractions; 32 PACs were identified from two of the several sub-fractions; 33 PACs were identified in a mutagenic, basic sub-fraction (Royer *et al.*, 1983).

Approximately 40 types of PAC were identified qualitatively in low-temperature reactor product or recycle oil from the HYGAS experimental gasifier. They included PAHs and nitrogen-, oxygen- and sulphur-substituted PACs (Stamoudis *et al.*, 1981).

Table 8. Constituents of coal-tars (average % by weight of dry tars)[a]

Component	Source of tars					
	Coke-oven			Vertical retort, UK	Low-temperature gasifier, UK	Lurgi gasifier, UK
	UK	FRG	USA			
Acenaphthene	0.96	0.3	1.05	0.66	0.19	0.57
Anthracene	1.00	1.8	0.75	0.26	0.06	0.32
Benzene	0.25	0.4	0.12	0.22	0.01	0.02
Carbazole	1.33	1.5	0.60	0.89	1.29	0.22
o-Cresol	0.32	0.2	0.25	1.33	1.48	1.14
m-Cresol	0.45	0.4	0.45	1.01	0.98	1.83
p-Cresol	0.27	0.2	0.27	0.86	0.87	1.51
Diphenylene oxide	1.50	1.4	-	0.68	0.19	0.57
Ethylbenzene	0.02	-	0.02	0.03	0.02	0.04
Fluorene	0.88	2.0	0.64	0.51	0.13	0.62
High-boiling tar acids	0.91	-	0.83	8.09	12.89	11.95
Medium-soft pitch	59.8	54.4	63.5	43.7	26.0	33.1
α-Methylnaphthalene	0.72	0.5	0.65	0.54	0.23	0.63
β-Methylnaphthalene	1.32	1.5	1.23	0.68	0.19	1.05
Naphtha	1.18	-	0.97	3.21	3.63	3.02
Naphthalene	8.94	10.0	8.80	3.18	0.65	2.01
Phenanthrene	6.30	5.7	2.66	1.75	1.60	0.28
Phenol	0.57	0.5	0.61	0.99	1.44	0.97
Styrene	0.04	-	0.02	0.04	0.01	0.01
Tar bases	1.77	0.73	2.08	2.09	2.09	2.50
Toluene	0.22	0.3	0.25	0.22	0.12	0.05
o-Xylene	0.04	-	0.04	0.06	0.05	0.05
m-Xylene	0.11	0.2	0.07	0.13	0.10	0.07
p-Xylene	0.04	-	0.03	0.05	0.04	0.03
Xylenols	0.48	-	0.36	3.08	6.36	5.55

[a]From McNeil (1983)

Concentrations of PACs, including polynuclear aromatic amines, were determined for tars produced in the Stoic fixed-bed gasifier at the University of Minnesota, USA (see Table 9) (Griest & Clark, 1982).

Tar from a liquid condensate produced by underground coal gasification was fractionated into neutral, acidic and basic fractions, and compounds or types of compounds from the basic fraction were identified. Substituted pyridines accounted for about 47% of the total basic fraction, anilines for 13% and quinolines for 9% (Huffstetler & Rickert, 1977).

Laboratory studies of coal distillation have demonstrated the effect of temperature, residence time at temperature, pressure, coal type, coal-particle heating rate, and the gaseous environment on the quantities, chemical distribution and properties of tars distilled under some simulated process conditions. After their distillation, the survival or cracking of the tars is dependent in large part on the nature of the gas flows in the specific process (Fillo *et al.*, 1983).

Table 9. Concentrations of polynuclear aromatic amines and hydrocarbons measured in the top of the Stoic fixed-bed gasifier at the University of Minnesota, USA[a]

Substance	Concentration (μg/g)
Polynuclear aromatic amines	
1-Aminonaphthalene	9
2-Aminonaphthalene	30
C_1-Aminonaphthalenes[b]	4-24[c]
C_2-Aminonaphthalenes[b]	7-10[d]
C_3-Aminonaphthalene[b]	7
Aminofluorene	9
1-Aminoanthracene	13
2-Aminoanthracene	2
C_1-Aminoanthracene[b]	1
Acridine	26
Polynuclear aromatic hydrocarbons	
Fluoranthene	180
Pyrene	110
Benzo[*a*]fluorene	150
Benzo[*b*]fluorene	100
3-Methylpyrene	90
Benz[*a*]anthracene	140
Chrysene/triphenylene	90
Benzo[*b*]-+-[*j*]fluoranthenes	140
Benzo[*a*]pyrene	80
Perylene	110

[a]From Griest and Clark (1982)

[b]Alkylated compound, e.g., C_2 means dimethyl or ethyl; C_3 means trimethyl, methyl ethyl or propyl

[c]Range of results for five separate isomers

[d]Range of results for three separate isomers

4. Biological Data Relevant to the Evaluation of Carcinogenic Risk to Humans

The studies of carcinogenicity and mutagenicity reported in this section include only those on samples of complex mixtures taken from the industry.

4.1 Carcinogenicity studies in animals

The literature dealing with the carcinogenicity of the components of tars is extensive, but investigations on the carcinogenicity of tars themselves are rather limited.

Skin application[1]

As early as 1913, a series of experiments on coal-tars [principally from gas-works] using rabbits was carried out, but no tumour was produced (Haga, 1913).

Mouse: Deelman (1923) showed that tar from horizontal retorts for gas production produced skin cancers in mice more rapidly than from vertical retorts. This difference was investigated by Kennaway (1925), who obtained three tars made from Durham Holmside coal heated at 450°C, 560°C, and 1250°C [the last of these figures would correspond to gas-works temperature]. These were painted on the dorsal skin of mice twice weekly and, by the 230th day of the experiment, skin tumours were found at the site of application in 10 of 20 surviving mice treated with tar from coal heated to 1250°C and in 10 of 29 surviving mice treated with tar from coal heated to 560°C, while only two of 49 surviving mice treated with tar from coal heated to 450°C developed tumours. Mortality was high subsequent to this date, and the experiment was terminated at day 373.

Because crude tars obtained from gas-works were usually toxic to mice when applied topically and led to the early death of most of the test group, Hieger (1929) investigated lower concentrations, to see if they would be equally effective in producing tumours. He prepared an ether extract of a gas-works tar and applied it twice weekly to the skin of three groups of 50 mice, either undiluted or as a 10% or a 1% solution in benzene. Average survival was 328, 341 and 512 days for the three groups, respectively. The incidence of skin tumours (most of which were carcinomas) was 22/50, 28/50 and 7/50, respectively, and the first tumours appeared at 63, 142 and 307 days.

Woglom and Herly (1929) studied the carcinogenic activity of a tar from a horizontal retort of a plant producing household gas, distilled at a temperature of about 1000°C. It was applied undiluted or as a 75%, a 50% or a 25% solution in glycerine topically to groups of 50 mice every second day on the interscapular region. When tar was applied in the undiluted form, the first skin carcinoma occurred after 163 days: 15 mice had survived until this day, and 10 of these developed skin tumours. With the 75%, 50% and 25% concentrations, 31, 27 and 30 mice, respectively, survived until the first appearance of the skin tumours at 138, 156 and 149 days. Among survivors, the numbers developing skin carcinomas were 10/31, 9/27 and 15/31, respectively.

The hypothesis that coal-tars contain several carcinogenic agents apart from benzo[*a*]pyrene was developed from the work of Berenblum (1945). Thus a horizontal-retort tar was obtained from the City of Leeds Gas Department (Berenblum & Schoental, 1947) and tested by topical application once weekly as a 5% solution in benzene on the skin of 20 mice. Of the 16 mice surviving for the 18-week duration of the experiment, eight developed skin tumours. The same tar was fractionated by chromatography on an aluminium oxide column. The benzene eluate produced skin tumours in 14/20 mice (14 mice had survived the 18 weeks of the experiment) but the adsorbates eluted with acetone or with a mixture of chloroform and acetic acid had no carcinogenic activity when tested in the same way. Chromatography of the benzene eluate with benzene and light petroleum yielded three fractions, which were tested in groups of 10 mice: a weakly adsorbed fraction produced skin tumours in 6/10 mice (nine surviving after 14 weeks of painting); a moderately adsorbed fraction gave skin tumours in 5/10 mice; and a strongly adsorbed fraction produced skin tumours in 1/10 mice. Chromatogra-

[1]The Working Group was aware of a study in progress to test the carcinogenicity of a coal-gasifier tar by skin application in mice (IARC, 1982).

phy of tar with a mixture of benzene and light petroleum yielded four fractions, which were also tested in groups of 10 mice. The first fraction consisted of material eluted after anthracene but before benzo[*a*]pyrene. When it was painted on the skin of mice once weekly for 16 weeks, three of 10 surviving mice developed skin tumours. The second fraction, containing benzo[*a*]pyrene, produced skin tumours in five of 10 surviving mice; while the two other fractions, eluted after benzo[*a*]pyrene, produced skin tumours in 6/10 (eight survivors) and 4/10 (seven survivors) mice, respectively, when applied in the same way. The tar was also extracted with light petroleum, and the soluble fraction, tested in a group of 10 mice, was found to induce skin tumours in 5/10 surviving mice after 15 weeks of treatment, while the insoluble fraction was inactive. The soluble fraction was chromatographed on an aluminium oxide column, and the subfractions were tested in groups of 20 mice. The first subfraction, eluted after anthracene but before benzo[*a*]pyrene, did not produce any skin tumours when painted on mice for 16 weeks. The second subfraction, containing benzo[*a*]pyrene, produced skin tumours in 10/20 mice (14 surviving), while the next fraction produced skin tumours in 5/20 (10 surviving). Two other subfractions were tested on the skin of mice but produced no skin tumours. The light-petroleum extract was chromatographed after acid/alkali treatment. The first fraction, obtained after anthracene but before benzo[*a*]pyrene, did not produce any skin tumours when tested on the skin of 11 mice for 28 weeks. The next fraction, which contained benzo[*a*]pyrene, produced skin tumours in seven of 10 mice tested (eight surviving) when tested for 20 weeks; while the fraction after this produced skin tumours in three of 10 mice tested (three surviving) when tested for 28 weeks. The strongly adsorbed residue did not produce skin tumours. The tar was also fractionated by vacuum distillation at different temperatures, from 120-180°C, and these fractions were tested in groups of varying numbers of mice. The first fraction was negative. The fraction that distilled at 140-160°C produced a skin tumour in one of 11 mice treated for 28 weeks (two surviving mice). The fractions obtained between 160-180°C and 180-200°C produced skin tumours in 4/10 (eight surviving) and 6/20 (four surviving) mice treated for 30 weeks. These last two fractions (weighing 80 g and 39 g) contained 50 mg and 240 mg of benzo[*a*]pyrene, respectively.

Crude tar obtained from a gas-coke works that used Pechora coal was tested for carcinogenicity after mixing with 14% benzene in order to soften its texture. The softened material was smeared on the interscapular skin of 13 strain A male mice three times weekly for six months (80 applications), and the animals were observed until they died. Seven mice survived to the first appearance of skin tumours. All surviving mice developed skin papillomas, and four developed carcinomas. In another experiment, 18 C_3H mice were treated with the tar mixed with 25% sunflower oil. Most of the mice died early on in the experiment from the toxic effect of the tar, and only five mice survived to the appearance of the first skin tumour. Two of these developed papillomas, and one developed a carcinoma. No carcinoma developed in 30 control mice of each strain (Grigorev, 1960).

Rabbit: Samples of a horizontal-retort tar obtained from the city of Leeds gas-works and fractions obtained from it, which were the same as those utilized in the experiment in mice (Berenblum & Schoental, 1947), were tested on rabbits twice weekly by repeated painting. Topical application of 5% tar in benzene produced skin tumours in six out of seven surviving rabbits after 14 weeks of treatment. The benzene eluate produced skin tumours in all six surviving rabbits treated for 14 weeks, but the adsorbates eluted with acetone or with a mixture of chloroform and acetic acid were inactive. The first and second fractions obtained from chromatography of the benzene eluate produced skin tumours in two of five surviving rabbits treated for 11 weeks; the third fraction produced no skin tumour. The first fraction obtained from the chromatographic separation of the tar in a mixture of benzene and light petroleum produced skin tumours in two of five rabbits (two surviving). The second fraction, containing benzo[*a*]pyrene, produced skin tumours in two of five rabbits (three surviving) after

15 weeks. The third and fourth fractions produced tumours in four of five (four surviving) and one of five (two surviving) rabbits, respectively. The soluble fraction of the light-petroleum extract of tar produced skin tumours in all five surviving rabbits treated for 14 weeks. The first subfraction produced skin tumours in four of five surviving rabbits treated for 16 weeks. The second subfraction (containing benzo[*a*]pyrene) produced skin tumours in three of five surviving rabbits treated for 16 weeks. The next three subfractions produced tumours in three of five, three of five and two of five surviving rabbits, respectively. The first fraction of the light-petroleum extract treated with acid and alkali produced skin tumours in all five surviving rabbits after 17 weeks of treatment. The same results were obtained with the second fraction (containing alkali- and acid-treated benzo[*a*]pyrene). The fraction after this produced skin tumours in four of five rabbits (four surviving rabbits), while the strongly adsorbed residue was inactive. The first vacuum-distilled fraction (120-140°C) was inactive; the second (140-160°C) produced skin tumours in two out of five (four surviving) rabbits; the third (160-180°C) and fourth (180-200°C) fractions each produced skin tumours in all five surviving rabbits (Berenblum & Schoental, 1947).

Five rabbits were treated with crude tar from a gas-coke works using Pechora coal, diluted in 14% benzene, three times weekly for six months (80 applications). All survived to the appearance of the first skin tumour; all developed papillomas, and four developed skin cancer. Ten rabbits were treated in the same way with tar mixed with 25% sunflower oil. All animals survived to the appearance of the first tumour; all developed papillomas, and one rabbit developed a carcinoma (Grigorev, 1960).

4.2 Other relevant biological data

(*a*) *Experimental systems*

Studies in this section refer to complex mixtures derived from modern gasification processes. No data were available on complex mixtures from older gas-works.

Toxic and reproductive effects

A number of toxicological and reproductive studies have been conducted on the residual ash and ash leachates (aqueous extracts) from the newer gasifiers, since these materials will be the major waste by-products requiring disposal and therefore represent a potential human and environmental health hazard.

Following Lurgi and similar coal-gasification processes, hot ash residue is removed at the bottom of the gasifier by a rotating grate. Several studies have examined the effects of ash added to the diet of rats (Blanusa *et al.*, 1979; Kostial *et al.*, 1979; Rabar *et al.*, 1979; Kostial *et al.*, 1980). Feeding of a diet containing 5% ash reduced body weight (Rabar *et al.*, 1979) and significantly reduced bone density (Blanusa *et al.*, 1979). This effect of ash on the bone was not potentiated by the presence of toxic metals in the drinking-water (cadmium, 100 mg/kg; manganese, 2000 mg/kg; or mercury, 50 mg/kg). These same investigators demonstrated that dietary treatment with ash (5%) had no effect on intestinal absorption or toxicity (measured by determining the LD_{50}) of these same toxic metals in suckling and adult albino rats (Kostial *et al.*, 1979).

Feeding of diets containing 0.5%, 1% and 5% ash for 16 weeks did not significantly affect urinary protein excretion, haematological parameters or iron, zinc and manganese concentrations in the kidneys, liver and testicles of albino rats (Kostial *et al.*, 1980).

Administration of E-effluent (water used for the quenching of ash) from a coal-gasification plant to albino rats as the drinking-water for nine months did not affect mortality rate, haematological parameters or concentrations of zinc, iron and manganese in the kidneys, liver, testicles and femur of rats. Femur composition and morphometry, gross pathology and organ histology were also unaffected (Kostial *et al.*, 1981). These same parameters were likewise unaffected by administration of a leachate (supernatant of ash) from a Lurgi gasifier, extracted in water for 24 hours and given to albino rats as the drinking-water for nine months (Kostial *et al.*, 1982).

Tar collected from a coal gasifier producing gas of a low calorific value, subfractions of the tar and some individual components were tested for cytotoxic activity in cultured dog alveolar macrophages *in vitro*. A subfraction, which contained phenols and neutral nitrogen heterocyclics, was the major contributor to the cytotoxicity of the tar sample (Hill *et al.*, 1983).

The effect of ash on reproduction in rats was reported by Rabar (1980). When both male and female rats were fed a diet containing 5% ash in a three-generation study, there was no effect on number of pregnancies or number of animals per litter. However, newborns had a slightly reduced body weight when compared with controls. A diet containing 1% ash had no effect on reproduction. No morphological or histopathological change was observed in newborns in response to maternal and paternal dietary treatment with ash.

Groups of four-month-old female and male albino rats (F_1 generation) were mated and exposed to gasifier-ash leachates as drinking-water *ad libitum*. Two subsequent generations of animals (F_2 and F_3) were also exposed to the same gasifier-ash leachates. Adequate control groups, which received tap water *ad libitum*, were also available. The experiment was terminated when the third generation litters were 14 days old. The numbers of pregnancies in the three successive generations of females exposed to ash leachates were slightly higher than in controls. No difference in the numbers of pups per litter or in the body weights of pups treated with ash leachates was observed when compared with the respective controls. No morphological or histopathological change was observed in the animals exposed to ash leachates. Concentrations of zinc, iron and manganese were measured in the carcasses of 14-day-old animals from the first and second experiments only. No difference between treated and control animals was observed (Kostial *et al.*, 1982). [The Working Group noted the incomplete reporting of this study.] The same parameters had been unaffected by administration of E-effluent in a three-generation study in rats (Kostial *et al.*, 1981).

Embryos of *Xenopus laevis* were exposed for 96 hours in aqueous extracts of tar from a coal gasifier (Schultz *et al.*, 1982). Mortality of the embryos increased with increasing concentrations of the tar extract. While the developmental stage of normal embryos was not affected by exposure to increasing concentrations of the extract, there was a reduction in embryo growth as a result of the exposure. Motility and pigmentation were both reduced. Various malformations were observed in the treatment groups, the severity of which was directly related to the concentration of tar extract to which the embryos were exposed.

Tests for genetic and related effects in experimental systems

No data from short-term genetic tests were available for tars or other samples taken from the older coal-gasification processes that were used to produce town gas. The following studies all utilized samples from new, in some cases experimental, gasification processes. In all studies, the metabolic activation system (S9) was derived from the livers of Aroclor-induced rats.

Process stream and waste by-product samples from several experimental coal gasifiers have been tested for mutagenicity in *Salmonella typhimurium*. An aqueous condensate (0.9% solids) from an experimental coal gasifier was mutagenic to strains TA98 and TA100 in the presence of S9 [no data were reported on activity in the absence of S9]. Extraction and fractionation of this condensate showed that most of the mutagenic activity for strain TA98 (with S9) was in the acidic and basic fractions (Epler *et al.*, 1978).

Samples from a pilot-scale, pressurized, fixed-bed gasifier (METC-stirred bed) have also been tested for mutagenicity in *S. typhimurium*. A tar from the disengagement chamber, tested at one dose, 250 μg, was mutagenic to strains TA1537, TA1538 and TA98 only in the presence of S9; some fractions of this tar were also mutagenic in the absence of S9 (Hobbs *et al.*, 1979). All the process-stream samples (raw gas, before Venturi scrubber, after Venturi scrubber, cleaned gas) were mutagenic to strains TA98 and TA100 with metabolic activation (Benson *et al.*, 1982a). The tar collected from the hot gas stream in a tar trap located immediately after an ash removal cyclone, was tested both as the whole tar and after fractionation in strain TA98 with and without metabolic activation. The unfractionated tar and most (10/12) of the fractions were found to be mutagenic to strain TA98 with metabolic activation. Most of the mutagenic activity was found in the polynuclear aromatic basic and neutral/phenolic fractions. No mutagenic activity was detected in the absence of S9 (Royer *et al.*, 1983).

Waste by-products (cyclone ash, bottom ash and scrubber water) from this same gasifier have also been tested for mutagenicity in *S. typhimurium* strain TA98. The ash samples were tested after extraction with dichloromethane (0.3% extractable). Six cyclone ash extracts were mutagenic to strain TA98 in the presence of S9. The bottom-ash extracts were very low in extractable organic compounds (0.01%) and were not mutagenic when tested at up to 100 μg per plate. The extract of water from the Venturi scrubber outlet (0.17% extractable) was mutagenic to strain TA98 in the presence of S9, and this mutagenicity was associated with the basic and neutral subfractions. The tar-trap tar was also mutagenic to strain TA98 but only with S9, and this mutagenicity was primarily in the fractions containing polynuclear aromatic hydrocarbons and their alkylated derivatives and in the fractions containing polar polynuclear compounds (Benson *et al.*, 1982b).

Process-stream samples from six locations in a coal-gasification pilot plant producing gas of a high-calorific value (the HYGAS, three-stage, fluidized-bed hydrogasification system; see Stamoudis *et al.*, 1981) were tested for mutagenicity in *S. typhimurium* strain TA98 in the presence and absence of S9. All of the process-stream samples, except the quench water from the pretreater, were mutagenic in the presence of S9 (Reilly *et al.*, 1981).

Three coal gasification by-products (a tar, cyclone dust and bottom ash) [from an unspecified type of gasifier] were tested in *S. typhimurium* strains TA1535, TA1538, TA98 and TA100. When suspended in dimethylsulphoxide (DMSO) and filtered prior to assay, only the gasification tar was mutagenic to TA1538, TA98 and TA100 and then only in the presence of S9. The tar was reported as negative in strain TA1535 [data not given]. The gasification particulate and ash were also reported to have no activity in strains TA98 and TA100; however, they were extracted only with DMSO and distilled water (pH 3-10) and were reported to be very insoluble. The gasification tar was extracted sequentially with a series of organic solvents; all of the extracts were mutagenic to *S. typhimurium* strain TA98 in the presence of S9 (Schoeny *et al.*, 1981a,b).

Crude tars collected from an experimental fixed-bed coal gasifier were mutagenic to *S. typhimurium* strain TA98 only in the presence of S9. The polynuclear aromatic, basic and polar neutral fractions isolated from the gasifier tar were all mutagenic to strain TA98 (in the presence of S9) when two different coals were used in the gasifier (Hughes *et al.*, 1983).

Environmental effluents from underground coal-gasification systems have been tested for mutagenicity in *S. typhimurium*. Two gasification sites were studied, one using air and the other oxygen to produce steam for injection into the gasifier. Dichloromethane extracts from groundwater, obtained from a network of sampling wells, were fractionated; both the neutral and basic fractions were mutagenic to strains TA98 and TA100 in the presence of S9. Further fractionation showed that most of the mutagenic activity was in the fraction containing aniline- and quinoline-related compounds. Tar condensed from the product-gas stream from these two underground gasifiers was mutagenic to strains TA98 and TA100 in the presence of S9. Fractionation of the tar showed that both the basic and neutral fractions were mutagenic to strains TA98 and TA100 in the presence of S9; however, most of the mutagenic activity was in the neutral fraction (Timourian *et al.*, 1982).

The basic fraction of tar condensed from the product-gas stream in an underground coal-gasification experiment induced mutations to thioguanine resistance and azaadenine resistance in a DNA repair-deficient cell line (UV-5) of Chinese hamster ovary (CHO) cells with metabolic activation from Aroclor-induced hamster-liver microsomes. The basic tar fraction also induced sister chromatid exchange in the CHO cells. At the same concentration range, neither the induction of chromosomal aberrations nor the induction of micronuclei was significantly increased (Timourian *et al.*, 1982). The authors also reported that the neutral fraction of the tar induced mutations, as measured by both types of drug resistance in the CHO cells in the presence of a microsomal supernatant of rat liver.

Different samples from the process stream (e.g., raw product gas, quench water) from the HYGAS coal-gasification pilot plant (see above) were reported to induce sister chromatid exchange in mouse myeloma cells (Reilly *et al.*, 1981). [No statistical evaluation of the data is presented.]

(*b*) *Effects in humans (other than cancer)*

The data on toxic effects in humans refer to workers in old town-gas plants.

Toxic effects

Doll *et al.* (1965) studied mortality from different causes in 11 499 gas workers over an eight-year period starting from September 1953 (see section 4.3). Occupational exposure was classified as high, intermittent or low. The differences were largely accounted for by bronchitis and lung cancer; death rates were highest in those most heavily exposed in the retorts and lowest in those with no exposure. Retort-house workers did not smoke more than other workers.

Doll *et al.* (1972) extended this study by a further four-year follow-up and concluded that the additional data offered only limited support to the view that bronchitis is a specific occupational hazard of gasworkers. The large excess of deaths from bronchitis observed earlier among retort-house workers was probably attributable to the working conditions prevalent in the 1940s.

4.3 Case reports and epidemiological studies of carcinogenicity in humans

(*a*) *Epidemiological studies on the destructive distillation of coal*

For over 200 years, skin cancers have been recognized to be associated with occupational exposure to tars and pitches related to the destructive distillation of coal. Studies from the early twentieth century drew attention to additional cancer sites, but they often did not distinguish the specific industry from which cases originated. It therefore seemed prudent to consider the early studies together.

Early authors alluded to case reports of scrotal cancer in previous work. O'Donovan (1920) reviewed the findings of Butlin (1892) and indicated that 34 of the 39 scrotal cancer cases admitted to St Bartholomew's Hospital, London, UK, were of industrial origin. The author described 16 additional skin cancers from exposure to tars and pitches. After the enactment in the UK of the Workmen's Compensation Act of 1907, 46 cases of scrotal cancer and 'epitheliomatous ulceration' were reported in 1907-1911 and 64 cases in 1914-1918. Of the latter cases, 44 were from two factories (O'Donovan, 1928). The later paper described the cases from the combined studies and found that, of the 33 cases for which a job was identified, nine were in tar workers.

Henry (1947) analysed 3753 cases of skin tumours that were reported to the Factory Inspectorate in the UK between 1920-1945. He attributed 2229 cases (59.4%) to tar and/or pitch exposure. The industries included tar distillery, patent fuel, as well as gas-works, coke ovens and coal carbonization. Scrotal lesions represented 21.3% of sites reported. The author noted that the time from first exposure to the time of diagnosis of a skin tumour was shorter for the tar and pitch products (20 years) than for oil exposures (50 years). [Compensation practices would have influenced the representation of various jobs and industries among the cases.]

Ross (1948) noted that in the UK the reported number of compensation cases of 'epitheliomatous ulceration' rose during 1920-1945 from 89 in the first five-year period to 449 in the last five years. The deaths, however, which initially were 15, had risen to 62 in 1931-1935 and fallen to nine in 1941-1945. The author pointed out that it is difficult to know whether these changes were due to variation in reporting practices or to a true increase in disease. Ross also reported the frequency of lesions by three-year periods from 1936-1947 in an average of 170 personnel employed in the manufacture of carbon products. The frequency of epithelioma ranged from 1.2-4.7% and that of papilloma from 10.6-23.5% with total numbers over the whole period of 16 and 102, respectively.

Henry *et al.* (1931) collected death certificates relating to 13 965 bladder and prostate cancer deaths in England and Wales for the period 1921-1928 and compared the observed distribution of occupations recorded on these certificates to those expected on the basis of the occupational groupings in the 1921 census. Although 11 429 male deaths from either cancer were included, the numbers of deaths among many of the individual job categories were few. [The consistency of the finding of elevated standardized mortality ratios (SMRs) for bladder cancer in nine of eleven occupational groups with exposure to coal gas, tars and pitches is noteworthy.] Patent-fuel labourers, tar-distillery workers and most job classifications from the gas-works had SMRs between 117 and 400. Only chimney sweeps and gas-works inspectors had ratios below 100. The SMR for bladder cancer in coke-oven workers was 122. The combined SMR can be calculated from the data provided by the authors: for these selected occupations with exposures to tars and pitches, the SMR is 144 (63/43.8), which is significant. There was no apparent association with a risk of prostate cancer in the same occupational groups in which a risk of bladder cancer occurred. A special study was made of 50 cases of bladder cancer with enquiry made of the work histories by interviews with work mates, spouses and employers; 44 were traced, and in 41 the occupational categorization was considered to be accurate.

Kennaway and Kennaway (1936, 1947) reported two studies using a similar method of analysis to that used by Henry *et al.* (1931), but selecting deaths from lung and laryngeal cancer in England and Wales and examining the distribution of occupations from the 1931 census. The deaths among males aged 20 years or more included 23 549 lung cancers and 14 869 laryngeal cancers in the total period 1921-1938. The number of lung cancer cases increased by 58% between the first 12-year period and the last six-year period, which would represent a real increase of 116%. The number of deaths from laryngeal cancer decreased slightly during this period. In both papers, the authors examined the developing epidemic of lung cancer in that period, in an attempt to relate that event to occupation. The investigation identified several occupations with exposure to tars and pitches which were associated with high mortality ratios (SMRs). The lung cancer SMRs were high for patent-fuel workers (571), gas stokers and coke-oven chargers (284), and several other occupations associated with gas or gas production, such as gas-works engine drivers, gas-works foremen, gas producers and gas fitters (138-202). The category of 'tar distillery and coke-oven workers' had a very low SMR (52), despite a risk of bladder cancer common to several of these groups. SMRs for laryngeal cancer were not high in the same occupational groups. Gas stokers and coke-oven chargers who had high mortality SMRs for lung cancer also had high SMRs for laryngeal cancer (213), but, among the occupational groups exposed in the gas-works, only engine drivers had a SMR for laryngeal cancer above 110 (227).

[The studies presented in the three papers above have common problems: even with deaths accumulated over an 18-year period, single categories of deaths were often so small that few deaths could have occurred from a specific cancer. Also, comparison is made between occupation stated on a death certificate and census data. The accuracy of diagnosis on a death certificate could be questioned, and no information is available on length or type of occupational exposure.]

(*b*) *Epidemiological studies on coal gasification workers*

Bruusgaard (1959) reported on 125 autopsies on 'gas workers', finding 12 tumours of the respiratory tract, five of the bladder, three of the oesophagus and 21 at other sites. He noted that the number of cancers of the respiratory tract was 'significantly higher than expected and so was the percentage of these cancers of the total number of cancers detected'. [Detailed justification for this statement is not apparent in this short report.]

The epidemiological reports available were limited to three analytical studies using data from the British gas industry (Doll, 1952; Doll *et al.*, 1965, 1972). The first study of one large gas company is expanded in the later studies to include additional years of exposure for four 'area boards' (companies). In the final report, attempts are made to establish work histories for the cancer cases. Comparisons are made between various levels of exposure to coal-gasification by-products within the industry and also to the mortality experience of England and Wales. Age-, class- and area-standardized rates are computed.

The 1952 investigation was restricted to a study of pensioned, non-salaried, male employees of a large London gas company. Pensions were available to men over 55 years of age, and the study included all those in receipt of a pension on 1 January 1939 or within 10 years of that date. Between 1939 and 1948, 840 deaths were registered for 11 162 man-years of study. Expected deaths were calculated for two external control groups: one from the age-, sex- and year-specific rates for England and Wales and the other from London rates. The figures for observed compared to expected cases were 840/831.5 for all causes and 25/10.4 for lung cancer ($p < 0.001$). The numbers of skin and bladder tumours were considered to be too small for reliable computation. Weighting by London rates lowered the ratio for lung cancer slightly (25/13.8; $p < 0.01$). When the population was divided into those who were employed inside the works and those who worked outside (as fitters, repair men and inspectors), lung cancer ratios were 17/8.6 ($p < 0.02$) and 8/5.2, respectively.

The second study (Doll *et al.*, 1965) was an eight-year follow-up of 11 499 men employed by four area gas boards from 1 September 1953. Criteria for inclusion in the study were at least five years of employment, age at entry into the study of 40-65 years, and having worked in the coal-carbonization plant. All but 50 men were traced (0.4% untraced). The 34 occupational categories were merged into four classes: coal-carbonization process worker (A), intermittent exposures in a coal-carbonization plant (B), by-product exposure (C_1), and no exposure (C_2). Comparison was made with the rates for England and Wales, as in the previous study. The SMRs for lung cancer were A, 169; B, 112; C_1 and C_2 combined (C), 100. For bladder tumours, the SMR for class A was 221 (5 cases; 14 cases for all classes combined); there was one scrotal tumour in class A. The ratios were statistically significant for trend across the three classes and elevated for each board in comparison with those for the relevant conurbation. An analysis by type of retort house showed an excess mortality from lung cancer among workers in horizontal-retort houses compared with those in vertical-retort houses (SMRs, 182/127 when expected numbers were based on the experience of other gas-workers). There was no evidence of an effect of smoking on the basis of a 10% sample of smoking habits in living workers; this was reinforced by the finding that bronchitis rates were similar in the two groups.

The latest report in Doll's studies of gas workers (Doll *et al.*, 1972) added four more years of follow-up to the previous study for classes A and C_1 and data from four additional gas boards for seven to eight years with 4687 men, all successfully traced. Nine new deaths were noted that should have been included in the previous study. The SMR for all causes for the four old boards, for the whole period 1953-1965, was 114 for class A and 80 for class C_1. For lung cancer, the SMRs were for A, 179 ($p < 0.001$) and for C_1, 75. The SMRs for bladder cancer were 235 ($p = 0.03$) for A and 76 for C_1. Three cases of scrotal cancer were also noted in class A. For the new boards added in this study, SMRs for class A were 134 for lung (not significant) and 153 for bladder (based on two cases). A case-control mortality study of retort-house workers only showed that all three dead workers with scrotal cancer and all 12 dead workers with bladder cancer had worked for longer periods as retort-house men, especially as top men or hydraulic-mains attendants. For lung cancer, there were 13 cases in top men or hydraulic mains attendants compared with eight who had worked elsewhere in the retort house.

Two additional points are worth noting. A contemporary environmental monitoring investigation by Lawther *et al.* (1965) noted a mean long-term concentration of benzo[*a*]pyrene of 3 μg/m³, and samples taken above an old horizontal-retort house had concentrations as high as 216 μg/m³. Concentrations of polynuclear aromatic hydrocarbons were usually higher in horizontal-retort houses than in vertical-retort houses. A study of gas-retort houses of a type similar to those in Doll's studies (Battye, 1966) showed the presence of 2-naphthylamine at a concentration of 2 ng/m³ in the air and 5.6 mg/kg in the crude-gas initial tar condensate. Lawther and Waller (1980) concluded that smoking was not a sufficient explanation for the increased risk of lung cancer.

Manz (1980) reported on the mortality experience of a cohort of seven occupational groups at gas and water municipal plants and at a private coking plant in Hamburg. The proportion of deaths from lung or bladder cancer among all deaths was reported to be higher in those groups exposed to coal, coke and tars compared with occupational groups without such exposure (office workers). [The Working Group noted the inadequate reporting and analysis of the results.]

5. Summary of Data Reported and Evaluation

5.1 Exposures

Town gas and industrial gases derived from the destructive distillation of coal are produced in thousands of plants throughout the world. The processes are based on several gasifier designs. Substantial exposures to airborne polynuclear aromatic compounds, together with concomitant exposure to a variety of other contaminants, have been measured in retort houses. Polynuclear aromatic compounds were also measured in airborne samples and identified in surface samples from newer gasification plants.

5.2 Experimental data

Crude coal-tars from several older gas-works have been tested for carcinogenicity by skin application in mice and rabbits. All tars produced a high yield of skin papillomas and carcinomas. Various experiments have shown that the carcinogenicity of such tars cannot be attributed solely to their content of benzo[*a*]pyrene.

Two studies suggested that tar from horizontal retorts was more active in producing skin tumours in mice than tar from vertical retorts.

All process-stream and waste by-product samples tested which contained newer gasifier tars were mutagenic to *Salmonella typhimurium*. Samples of bottom ash from gasification systems did not show mutagenic activity. Several studies showed that mutagenic activity is found not only in fractions containing polynuclear aromatic compounds, but also in the more polar neutral, basic and total neutral fractions. The basic fraction of tar from one underground coal-gasification process was mutagenic and induced sister chromatid exchange in mammalian cells *in vitro*. No data on cell transformation were available.

5.3 Human data

Chronic bronchitis was reported to have occurred among employees in older gas-works.

Case reports of tumours of the skin (including the scrotum), bladder and respiratory tract in association with employment in industries involving the destructive distillation of coal

Overall assessment of short-term tests on tars and tar fractions from experimental gasifiers[a]

	Genetic activity			Cell transformation
	DNA damage	Mutation	Chromosomal effects	
Prokaryotes		+[b,c]		
Fungi/ Green plants				
Insects				
Mammalian cells (*in vitro*)		+[b]	+[b]	
Mammals (*in vivo*)				
Humans (*in vivo*)				
Degree of evidence in short-term tests for genetic activity: *Sufficient* for underground gasification tars *Inadequate* for other tars				Cell transformation: No data

The groups into which the table is divided and '+', '–' and '?' are defined on pp. 16-17 of the Preamble; the degrees of evidence are defined on pp. 17-18.

[b]The basic fraction of a tar from an underground coal-gasification system

[c]All tar samples were positive (bottom ash and bottom-ash leachate samples were negative)

suggested a link between that industry and human cancer. Despite their methodological shortcomings, the descriptive epidemiological studies based on death certificates corroborated these early suggestions.

A series of detailed analytical epidemiological studies of the British gas industry add further weight to the hypothesis that work in such coal gasification plants carries a risk of tumours of the lung, bladder and scrotum. Notwithstanding the limited details available on the work histories of the gas workers, there appeared to be a relationship between elevated relative risk of tumours and work in retort houses, particularly when the job entailed exposure to fume emanating from the retorts.

5.4 Evaluation[1]

There is *sufficient evidence* from observations made in the first half of this century that exposures to coal-tar from the destructive distillation of coal give rise to skin cancer in humans.

The available epidemiological studies provide *sufficient evidence* that certain exposures in the retort houses of the older coal-gasification processes are carcinogenic to humans, giving rise to lung cancer. The likely causative agent is coal-tar fume. There is *limited evidence* that such occupational exposures produce bladder cancer.

[1]For definitions of the italicized terms, see Preamble, pp. 15 and 19.

There is *sufficient evidence* that topical application of coal-tars from several older gas-works on the skin of experimental animals produced a high yield of skin cancers.

A number of individual polynuclear aromatic compounds for which there is *sufficient evidence* of carcinogenicity in experimental animals have been measured at high levels in air and tar samples taken from certain areas in coal-gasification plants[2].

The available evidence indicates that certain exposures in retort houses of older coal-gasification processes are carcinogenic to humans. No data on carcinogenicity in experimental animals or in humans from exposures during the newer processes were available to the Working Group.

6. References

Battye, R. (1966) *Bladder carcinogens occurring during the production of 'town' gas by coal carbonisation*. In: *Hygiene Toxicology. Occupational Diseases, Proceedings of the 15th International Congress on Occupational Health, Vienna, 19-24 Sept. 1966*, Vol. 3, Vienna, H. Egermann, pp. 153-158

Benson, J.M., Hill, J.O., Mitchell, C.E., Newton, G.J. & Carpenter, R.L. (1982a) Toxicological characterization of the process stream from an experimental low Btu coal gasifier. *Arch. environ. Contam. Toxicol.*, *11*, 363-371

Benson, J.M., Mitchell, C.E., Royer, R.E., Clark, C.R., Carpenter, R.L. & Newton, G.J. (1982b) Mutagenicity of potential effluents from an experimental low Btu coal gasifier. *Arch. environ. Contam. Toxicol.*, *11*, 547-551

Berenblum, I. (1945) 9,10-Dimethyl-1,2-benzanthracene as a highly potent carcinogen for the rabbit's skin. *Cancer Res.*, *5*, 265-268

Berenblum I. & Schoental, R. (1947) Carcinogenic constituents of coal-tar. *Br. J. Cancer*, *1*, 157-165

Blanuša, M., Kostial, K., Matković, V. & Landeka, M. (1979) Cortical index of the femur in rats exposed to some toxic metals and ash from coal gasification. *Arh. Hig. Rada Toksikol.*, *30* (*suppl.*), 335-340

Braunstein, H.M., Copenhaver, E.D. & Pfuderer, H.A., eds (1977) *Environmental, Health and Control Aspects of Coal Conversion: An Information Overview*, Vol. 1, Oak Ridge, TN, Oak Ridge National Laboratory

Bruusgaard, A. (1959) Occurrence of certain forms of cancer in gasworkers (Norw.). *T. Norsk. Laegefor.*, *79*, 755-756

---

[2]See Appendix, Table 2.

Butlin, H.T. (1892) Three lectures on cancer of the scrotum in chimney-sweep and others. *Br. med. J.*, *2*, 1341-1346; *3*, 1-6; *3*, 66-71

Commins, B.T. (1958) Polycyclic hydrocarbons in rural and urban air. *Int. J. Air Pollut.*, *1*, 14-17

Cubit, D.A. & Tanita, R.K. (1982) *Industrial Hygiene Assessment of Coal Gasification Plants* (*NIOSH Contract No. 210-78-0040, Morgantown, WV*), National Institute for Occupational Safety & Health

Dailey, N.S. (1977) *Process effluents: Quantities and control technologies*. In: Braunstein, H.M., Copenhaver, E.D. & Pfuderer, H.A., eds, *Environmental, Health and Control Aspects of Coal Conversion: An Information Overview*, Oak Ridge, TN, Oak Ridge National Laboratory, pp. 4-1 - 4-157

Deelman, H.-T. (1923) Some remarks on experimental tar (Fr.). *Bull. Assoc. fr. Etude Cancer*, *12*, 24-30

Doll, R. (1952) The causes of death among gas-workers with special reference to cancer of the lung. *Br. J. ind. Med.*, *9*, 180-185

Doll, R., Fisher, R.E.W., Gammon, E.J., Gunn, W., Hughes, G.O., Tyrer, F.H. & Wilson, W. (1965) Mortality of gasworkers with special reference to cancers of the lung and bladder, chronic bronchitis, and pneumoconiosis. *Br. J. ind. Med.*, *22*, 1-12

Doll, R., Vessey, M.P., Beasley, R.W.R., Buckley, A.R., Fear, E.C., Fisher, R.E.W., Gammon, E.J., Gunn, W., Hughes, G.O., Lee, K. & Norman-Smith, B. (1972) Mortality of gasworkers - Final report of a prospective study. *Br. J. ind. Med.*, *29*, 394-406

Electric Power Research Institute (1983) *Coal Gasification Systems: A Guide to Status, Applications and Economics* (*EPRI AP-3109*), Palo Alto, CA, pp. 3-1, 3-2, 5-1

Epler, J.L., Larimer, F.W., Rao, T.K., Nix, C.E. & Ho, T. (1978) Energy-related pollutants in the environment: Use of short-term tests for mutagenicity in the isolation and identification of biohazards. *Environ. Health Perspect.*, *27*, 11-20

Fillo, J.P., Stamoudis, V.C., Stettler, J.R. & Vance, S.W. (1983) *Influence of Coal Devolatilization Conditions on the Yield, Chemistry and Toxicology of By-product Oils and Tars* (*DOE/ET/14746-11, ANL/Ser-2*), Morgantown, WV, Morgantown Energy Technology Center

von Fredersdorff, C.G. & Elliott, M.A. (1963) *Coal gasification*. In: Lowry, H.H., ed., *Chemistry of Coal Utilization*, Supplementary Volume, New York, John Wiley & Sons, pp. 893, 954

Griest, W.H. & Clark, B.R. (1982) *Sample Management and Chemical/Physical Properties of Products and Effluents from the University of Minnesota-Duluth, Low-BTU Gasifier* (*Contract No. W-7405-eng-26, ORNL/TM-8427*), Oak Ridge, TN, Oak Ridge National Laboratory

Grigorev, Z.E. (1960) On the carcinogenic properties of the Pechora coal tar. *Prob. Oncol.*, *6*, 883-886

Haga, I. (1913) Experimental studies on the occurrence of atypical epithelial and mucous skin proliferations (Ger.). *Z. Krebsforsch.*, *12*, 525-576

Henry, S.A. (1947) Occupational cutaneous cancer attributable to certain chemicals in industry. *Br. med. Bull.*, *4*, 389-401

Henry, S.A., Kennaway, N.M. & Kennaway, E.L. (1931) The incidence of cancer of the bladder and prostate in certain occupations. *J. Hyg.*, *31*, 125-137

Hieger, I. (1929) The influence of dilution on the carcinogenic effect of tar. *J. Pathol. Bacteriol.*, *32*, 419-423

Hill, J.O., Royer, R.E., Mitchell, C.E. & Dutcher, J.S. (1983) *In vitro* cytotoxicity to alveolar macrophages of tar from a low-Btu coal gasifier. *Environ. Res.*, *31*, 484-492

Hobbs, C.H., McClellan, R.O., Clark, C.R., Henderson, R.F., Griffis, L.C., Hill, J.O. & Royer, R.E. (1979) *Inhalation toxicology of primary effluents from fossil fuel conversion and use.* In: *Proceedings of the Symposium on Potential Health and Environmental Effects of Synthetic Fossil Fuel Technologies* (*CONF-780903*), Oak Ridge, TN, Oak Ridge National Laboratory, pp. 163-175

van der Hoeven, B.J.C. (1945) *Producers and producer gas.* In: Lowry, H.H., ed., *Chemistry of Coal Utilization*, Vol. 2, New York, John Wiley & Sons, pp. 1587, 1616

Huebler, J. & Janka, J.C. (1980) *Fuels, synthetic (gaseous)*, In: Mark, H.F., Othmer, D.F., Overberger, C.G. & Seaborg, G.T., eds, *Kirk-Othmer Encyclopedia of Chemical Technology*, Vol. 11, New York, John Wiley & Sons, pp. 410-446

Huffstetler, J.K. & Rickert, L.W. (1977) *Conversion processes.* In: Braunstein, H.M., Copenhaver, E.D. & Pfuderer, H.A., eds, *Environmental, Health and Control Aspects of Coal Conversion: An Information Overview*, Oak Ridge, TN, Oak Ridge National Laboratory, pp. 3-1-3-125

Hughes, T.J., Wolff, T.J., Nicols, D., Sparacino, C. & Kolber, A.R. (1983) *Synergism and antagonism in complex environmental mixtures: Unmasking of latent mutagenicity by chemical fractionation.* In: Kolber, A., Wong, T.K., Grant, L.D., DeWoskin, R.S. & Hughes, T.J., eds, In Vitro *Toxicity Testing of Environmental Agents*, Part B, Oxford, Plenum, pp. 115-144

IARC (1982) *Information Bulletin on the Survey of Chemicals Being Tested for Carcinogenicity*, No. 10, Lyon, p. 122

Kennaway, E.L. (1925) Experiments on cancer producing substances. *Br. med. J.*, *ii*, 1-4

Kennaway, E.L. & Kennaway, N.M. (1947) A further study of the incidence of cancer of the lung and larynx. *Br. J. Cancer*, *1*, 260-298

Kennaway, N.M. & Kennaway, E.L. (1936) A study of the incidence of cancer of the lung and larynx. *J. Hyg.*, *36*, 236-267

Kostial, K., Kello, D., Rabar, J., Maljković, T. & Blanuša, M. (1979) Influence of ash from coal gasification on the pharmacokinetics and toxicity of cadmium, manganese and mercury in suckling and adult rats. *Arh. Hig. Rada Toksikol.*, *30 (suppl)*, 319-326

Kostial, K., Kello, D., Blanuša, M., Maljković, T., Rabar, I., Bunarević, A. & Stara, J.F. (1980) Toxicologic studies of emissions from coal gasification process. I. Subchronic feeding studies. *J. environ. Pathol. Toxicol.*, *4*, 437-448

Kostial, K., Blanuša, M., Rabar, I., Maljković, T., Kello, D., Landeka, M., Bunarević, A. & Stara, J.F. (1981) Chronic studies in rats exposed to liquid effluent from coal gasification process. *J. appl. Toxicol.*, *1*, 3-10

Kostial, K., Rabar, I., Blanuša, M., Kello, D., Maljković, T., Landeka, M., Bunarević, A. & Stara, J.F. (1982) Chronic and reproduction studies in rats exposed to gasifier ash leachates. *Sci. total Environ.*, *22*, 133-147

Kreyberg, L. (1959) 3:4-Benzpyrene in industrial air pollution: Some reflexions. *Br. J. Cancer*, *13*, 618-622

Lamey, S.C., McCaskill, K.B. & Smith, R.R. (1981) *GC/MS Characterization of Condensable Tars in the Output Stream of a Stirred Fixed-Bed Gasifier (DOE/METC/TPR-82/2 (DE 82011723))*, Morgantown, WV, United States Department of Energy, Morgantown Energy Technology Center

Lawther, P.J. & Waller, R.E. (1980) *Lung cancer mortality among gas workers*. In: Werner, W. & Schneider, H.W., eds, *Air Contamination by Polycyclic Aromatic Hydrocarbons* (Ger.) (*VDI-Berichte No. 358*), Düsseldorf, Vereins Deutscher Ingenieure, pp. 223-226

Lawther, P.J., Commins, B.T. & Waller, R.E. (1965) A study of the concentrations of polycyclic aromatic hydrocarbons in gas works retort houses. *J. ind. Med.*, *22*, 13-20

Lindstedt, G. & Sollenberg, J. (1982) Polycyclic aromatic hydrocarbons in the occupational environment with special reference to benzo[*a*]pyrene measurements in Swedish industry. *Scand. J. Work Environ. Health*, *8*, 1-19

Lowry, H.H., ed. (1945) *Chemistry of Coal Utilization*, Vol. 2, New York, John Wiley & Sons, pp. 1587-1616

Lowry, H.H., ed. (1963) *Chemistry of Coal Utilization*, Supplementary Volume, New York, John Wiley & Sons, pp. 893, 954

Manz, A. (1980) *Respiratory and urinary tract as indicator of occupational (coal tar) carcinomas in carbonization and pipe workers*. In: Werner, W. & Schneider, H.W., eds, *Air Contamination by Polycyclic Aromatic Hydrocarbons* (Ger.) (*VDI-Berichte No. 358*), Düsseldorf, Vereins Deutscher Ingenieure, pp. 227-235

McNeil, D. (1983) *Tar and pitch*. In: Mark, H.F., Othmer, D.F., Overberger, C.G. & Seaborg, G.T., eds, *Kirk-Othmer Encyclopedia of Chemical Technology*, 3rd ed., Vol. 22, New York, John Wiley & Sons, pp. 564-600

National Institute for Occupational Safety and Health (1978) *Occupational Exposures in Coal Gasification Plants* (*DHEW (NIOSH) Publ. No. 78-191*), Cincinnati, OH

O'Donovan, W.J. (1920) Epitheliomatous ulceration among tar workers. *Br. J. Dermatol. Syph.*, *32*, 215-228

O'Donovan, W.J. (1928) *Cancer of the skin due to occupation*. In: *International Conference on Cancer*, London, Fowler Wright Ltd, pp. 292-303

Ondich, G.G. (1983) *Pollution Control Technical Manual for Lurgi-Based Indirect Coal Liquefaction and SNG* (*EPA-600/8-83-006*), Washington DC, US Environmental Protection Agency, pp. 14-80, 134

Parekh, R.D. (1982) *Handbook of Gasifiers and Gas Treatment Systems* (*Contract No. DE-ACO1-78ET10159*), Washington DC, US Department of Energy

Perch, M. (1979) *Coal conversion processes (carbonization)*. In: Mark, H.F., Othmer, D.F., Overberger, C.G. & Seaborg, G.T., eds, *Kirk-Othmer Encyclopedia of Chemical Technology*, 3rd ed., Vol. 6, New York, John Wiley & Sons, pp. 284-306

Rabar, I. (1980) The effect of ash from coal gasification on reproduction in rats (Abstract No. P. 259). *Toxicol. Lett., Special Issue 1*, 257

Rabar, I., Maljković, T., Kostial, K. & Bunarević, A. (1979) The influence of ash from coal gasification on body weight in relation to age and sex. *Arh. Hig. Rada Toksikol., 30 (suppl.)*, 327-334

Reilly, C.A., Jr, Peak, M.J., Matsushita, T., Kirchner, F.R. & Haugen, D.A. (1981) *Chemical and biological characterization of high-Btu coal gasification (the HYGAS process). IV. Biological activity*. In: Mahlum, D.D., Gray, R.H. & Felix, W.D., eds, *Coal Conversion and the Environment: Chemical, Biomedical and Ecological Considerations. 20th Annual Hanford Life Sciences Symposium, Richland, Wash., Oct. 1980* (*CONF-801039*), Washington DC, Technical Information Center, US Department of Energy, pp. 310-324

Ross, P. (1948) Occupational skin lesions due to pitch and tar. *Br. med. J., ii*, 369-374

Royer, R.E., Mitchell, C.E., Hanson, R.L., Dutcher, J.S. & Bechtold, W.E. (1983) Fractionation, chemical analysis, and mutagenicity testing of low-Btu coal gasifier tar. *Environ. Res., 31*, 460-471

Schoeny, R., Warshawsky, D., Hollingsworth, L., Hund, M. & Moore, G. (1981a) Mutagenicity of coal gasification and liquifaction products. *Environ. Sci. Res., 22*, 461-475

Schoeny, R., Warshawsky, D., Hollingsworth, L., Hund, M. & Moore, G. (1981b) Mutagenicity of products from coal gasification and liquifaction in the *Salmonella*/microsome assay. *Environ. Mutagenesis, 3*, 181-195

Schultz, T.W., Dumont, J.N., Clark, B.R. & Buchanan, M.V. (1982) Embryotoxic and teratogenic effects of aqueous extracts of tar from a coal gasification electrostatic precipitator. *Teratog. Carcinog. Mutagenesis, 2*, 1-11

Stamoudis, V.C., Bourne, S., Hangen, D.A., Peak, M.J., Reilly, C.A., Stetter, J.R. & Wilzback, K. (1981) *Chemical and biological characterization of high-Btu coal gasification (the HYGAS process). 1. Chemical characterization of mutagenic fractions*. In: Mahlum, D.D., Gray, R.H. & Felix, W.D., eds, *Coal Conversion and the Environment: Chemical, Biomedical and Ecological Considerations. 20th Annual Hanford Life Sciences Symposium, Richland, Wash., Oct. 1980* (*CONF-801039*), Washington DC, Technical Information Center, US Department of Energy, pp. 67-95

Timourian, H., Felton, J.S., Stuermer, D.H., Healy, S., Berry, P., Tompkins, M., Battaglia, G., Hatch, F.T., Thompson, L.H., Carrano, A.V., Minkler, J. & Salazar, E. (1982) Mutagenic and toxic activity of environmental effluents from underground coal gasification experiments. *J. Toxicol. environ. Health*, *9*, 975-994

Vorres, K.S. (1979) *Coal*. In: Mark, H.F., Othmer, D.F., Overberger, C.G. & Seaborg, G.T., eds, *Kirk-Othmer Encyclopedia of Chemical Technology*, 3rd ed., Vol. 6, New York, John Wiley & Sons, pp. 224-283

Williams, T.I. (1981) *A History of the British Gas Industry*, Oxford, Oxford University Press

Woglom, W.H. & Herly, L. (1929) The carcinogenic activity of tar in various dilutions. *J. Cancer Res.*, *13*, 367-372

Young, R.J. & Evans, J.M. (1979) Occupational health and coal conversion. *Occup. Health Saf.*, *48*, 22-26

Young, R.J., McKay, W.J. & Evans, J.M. (1978) Coal gasification and occupational health. *Am. ind. Hyg. Assoc. J.*, *39*, 985-997

COKE PRODUCTION

Coke production, as described in this monograph, is confined principally to slot-oven coke batteries, and includes all steps from coal handling to the sizing of the finished coke.

1. Historical Perspectives

Much of the information in this section was obtained from McGannon (1971).

It has not been definitely determined when man first learned to produce iron from its various ores. There have been suggestions that this could have been accomplished as early as 2000 BC. However, it is generally believed that the reduction of iron ore and the resultant production of iron became a widespread practice between 1350 and 1100 BC. A process used for many centuries was one in which iron ore, embedded into burning charcoal, produced metallic iron. The charcoal served as both the source of heat and the reducing agent for the iron oxide. The charcoal also protected the reduced iron from being reoxidized by oxygen from the air, and the carbon was absorbed by the hot, reduced iron to form various iron-carbon alloys.

Charcoal was used as both the fuel and reducing agent until the 1700s. Coke, which is the residue from the destructive distillation of coal, was first used as a blast-furnace fuel in England in 1619, but its use was not adopted generally until the mid-1700s. Coke was produced in two types of oven. The first oven constructed was the 'beehive', a term which accurately describes its shape. The coal was charged into the top of the oven from a hopper on wheels, or larry car, and then levelled either by hand or machine. The door on the side of the oven was then completely bricked. After the coal was coked, the brickwork was removed and the coke was sprayed with water to cool it and then removed by machine or hand. In this type of oven, the volatiles were not collected and the emissions were allowed to escape into the atmosphere.

The second type of oven is the by-product, or slot oven. In this type of oven, the volatilized products are recovered. The development of the by-product oven began in the early 1800s when gas was made from coal. By the 1850s a coke oven had been developed that could produce coke and recover the volatiles. Refinements were made to the heating system during the next few decades. In 1895, by-product ovens were installed for the first time in a US steel plant for the production of coke. Today, the vast majority of coke is produced in this type of oven.

2. Description of the Industry

This subject was reviewed by McGannon (1971) and the US Occupational Safety and Health Administration (1976).

2.1 Coal characteristics

Bituminous coal is used in the coking process. High-volatile bituminous coal is usually blended with low-volatile and/or medium-volatile coal. This blending of the different coals provides a coke of sufficient quality and strength. The coals should also contain minimal

Table 1. Mean analytical values for 101 bituminous coals[a]

Constitutent	Mean	Minimum	Maximum
Arsenic	14.02 mg/kg	0.50	93.00
Boron	102.21 mg/kg	5.00	224.00
Beryllium	1.61 mg/kg	0.20	4.00
Bromine	15.42 mg/kg	4.00	52.00
Cadmium	2.52 mg/kg	0.10	65.00
Cobalt	9.57 mg/kg	1.00	43.00
Chromium	13.75 mg/kg	4.00	54.00
Copper	15.16 mg/kg	5.00	61.00
Fluorine	60.94 mg/kg	25.00	143.00
Gallium	3.12 mg/kg	1.10	7.50
Germanium	6.59 mg/kg	1.00	43.00
Mercury	0.20 mg/kg	0.02	1.60
Manganese	49.40 mg/kg	6.00	181.00
Molybdenum	7.54 mg/kg	1.00	30.00
Nickel	21.07 mg/kg	3.00	80.00
Palladium	71.10 mg/kg	5.00	400.00
Lead	34.78 mg/kg	4.00	218.00
Antimony	1.26 mg/kg	0.20	8.90
Selenium	2.08 mg/kg	0.45	7.70
Tin	4.79 mg/kg	1.00	51.00
Vanadium	32.71 mg/kg	11.00	78.00
Zinc	272.29 mg/kg	6.00	5350.00
Zirconium	72.46 mg/kg	8.00	133.00
Aluminium	1.29 %	0.43	3.04
Calcium	0.77 %	0.05	2.67
Chlorine	0.14 %	0.01	0.54
Iron	1.92 %	0.34	4.32
Potassium	0.16 %	0.02	0.43
Magnesium	0.05 %	0.01	0.25
Sodium	0.05 %	0.00	0.20
Silicon	2.49 %	0.58	6.09
Titanium	0.07 %	0.02	0.15
Organic sulphur	1.41 %	0.31	3.09
Pyritic sulphur	1.76 %	0.06	3.78
Sulphate sulphur	0.10 %	0.01	1.06
Total sulphur	3.27 %	0.42	6.47
Moisture	9.05 %	0.01	20.70
Volatile matter	39.70 %	18.90	52.70
Fixed carbon	48.82 %	34.60	65.40
Ash	11.44 %	2.20	25.80
Carbon	70.28 %	55.23	80.14
Hydrogen	4.95 %	4.03	5.79
Nitrogen	1.30 %	0.78	1.84
Oxygen	8.68 %	4.15	16.03
High-temperature ash	11.41 %	3.28	25.85
Low-temperature ash	15.28 %	3.82	31.70

[a]From National Insitute for Occupational Safety and Health (1978)

amounts of both sulphur and ash, as these remain in the coke and, in the blast furnace, increase the volume of slag produced, increase coke use, decrease the production of iron, and make it more difficult to control the furnace. The chemical and structural properties, size, strength to withstand breakage and abrasion, bulk density, and combustibility are the most important features of good quality metallurgical coke. Table 1 gives the mean analytical values for 101 bituminous coals.

In the USA, the primary product recovered from the coking process is coke, while in Europe, the recovery of gas is as important as the coke. This has led to slight differences in the operation of the two processes. In the USA, the coal has a lower moisture content, the volatile matter is greater, and higher temperatures and faster coking rates are used. In Europe, the coal is crushed to a greater degree, the bulk density is lower, and the coke yields are higher, but the yields of tars and light oils are lower (Doherty & De Carlo, 1967). However, the structure and function of the ovens are the same.

2.2 Structure of coke ovens

A by-product coke battery usually consists of 10-100 ovens arranged in parallel rows. A coke plant comprises from 1-12 batteries and also includes the coal-unloading station, a system of conveyors to transfer the coal to a storage bunker; a coke-quenching station, where the coke is cooled; a coke wharf, where the quenched coke is stored; a coke-screening station, where the coke is sized; and, finally, a car-loading station (Fig. 1).

Fig. 1. View of a typical coke production plant

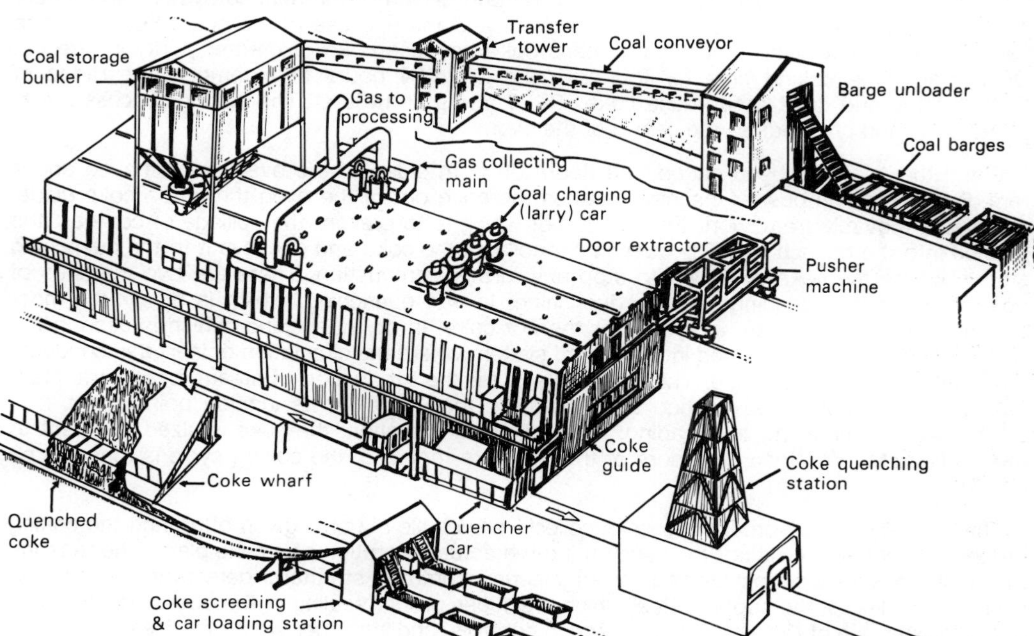

The individual by-product or slot oven is basically composed of the coking chamber, heating flues and a regenerative chamber. The heating flues and coking chambers alternate such that a heating chamber is located on each side of a coking chamber. The regenerative chamber is located beneath the other two. The coking chamber is the place in which the coal is destructively distilled into coke, the heating flues or combustion chambers are where the fuel gas is burned, and the regenerative chambers contain the regenerators that permit separate control of the flow of preheated air for combustion to the individual, vertical heating-flue walls. The arrangement of individual regenerators allows for the close individual control of the heating of each oven. There are doors on both ends of the refractory-bricked coking chamber. The ends are designated as the 'push side', where the pushing ram enters to force the coked coal out of the oven, and the 'coke side'. The most common oven is 3-6 m in height, 11-15 m in length, and 42.5-50 cm in width. The newer ovens are larger; their dimensions are approximately 7 m in height, 15 m in length, and 45 cm in width.

2.3 Operating characteristics

The coking cycle begins when the coal is charged or dropped into the coking chamber through holes on top of the oven. A larry car or mechanical device, which has hoppers for the coal and travels on rails, brings the coal to the charging holes. There are usually three or four charging holes per oven. Before the ovens are charged, the lids that cover the charging holes are removed, and the oven is then placed on steam aspiration to prevent the escape of gases into the atmosphere. After the coal charge has been dropped into the oven, the lids are replaced and luted, if necessary. Luting is a process in which the seal of the lid to the hole is coated with refractory material, a procedure which helps to prevent the escape of emissions into the atmosphere. The aspiration system is also shut off.

The coking time is a function of the condition and design of the heating system, the nature of the coals to be coked, the width of the oven, and the end use of the coke. Foundry coke has a longer coking time than blast-furnace coke because coke of greater purity is needed. On average, the coking time for foundry coke is 25-34 hours at a temperature range of 1000-1150°C and, for blast-furnace coke, 16-20 hours at 1400°C. The coking process starts at the walls and proceeds to the centre of the oven.

When the coal has been coked, the doors on both sides of the oven are removed and a ram, located on the pusher machine, pushes the coke out of the oven through a coke guide, which is a movable framework the function of which is to prevent the spillage of coke as it is pushed into the quench car. The quench car collects the coke and conveys it to the quenching station where the coke is cooled to stop any further combustion. There are two methods of doing this - wet quenching and dry quenching. In wet quenching, which is by far the most common method, water is sprayed over the hot coke. In dry quenching, which is used in the USSR, the hot coke is dumped into a closed system where the circulation of inert gas conducts the heat away from the coke. This heat is used for the production of steam for general plant use. When the coke has reached a temperature that is below its ignition point in air, it is discharged for screening and loading. At the screening station, the coke is sized and sent to the loading station. When the doors of the oven are replaced, the coking cycle is ready to be repeated.

Theoretically, the by-product process collects all volatile material given off during the coking process. However, this is not the case during the actual operation of a coke plant. The reasons that emissions escape include lack of engineering controls, structural defects in the battery, such as buckling of the support steel which precludes proper sealing of the doors and charging lids, improper use of the installed engineering controls, and improper work practices.

present during the production of coke. The sampling method and analytical technique for both CTPV and coke-oven emissions are the same today, except that the extraction method from the sampling media for coke-oven emissions is ultrasonic and that for CTPV is a Soxhlet extraction. The results, whether they are designated CTPV or coke-oven emissions, are comparable. Fannick *et al.* (1972) analysed 319 samples collected in the workers' breathing zone, and the results are summarized in Table 4.

Table 4. Exposure to the benzene-soluble fraction of coke-oven particulates according to job title[a]

Job title	Number of samples	Concentration (mg/m³)		
		Average	Median	Range
Coke-side benchman	18	1.08	0.77	0.09-2.74
Door-machine operator	25	2.11	1.70	0.04-6.51
Heater	39	1.07	0.82	ND[b]-2.98
Larry-car operator	39	3.05	2.73	0.28-8.78
Lidman	61	3.22	2.30	0.42-17.89
Luterman	18	2.57	2.30	0.25-4.82
Miscellaneous	18	0.93	0.95	0.18-2.76
Pusher-machine operator	23	0.39	0.33	ND-0.93
Pusher-side benchman	28	2.03	1.17	ND-14.59
Quench-car operator	23	0.94	0.43	ND-7.20
Tar chaser	27	3.14	2.37	0.04-14.70
	319 (total)	2.08 (mean)	1.42 (mean)	ND-17.89 (total range)

[a]From Fannick *et al.* (1972)

[b]ND, none detected

The National Institute for Occupational Safety and Health (1975) in the USA performed sampling to determine the effectiveness of coke-side sheds for controlling emissions. The mean exposure to CTPV under a coke-side shed was 1.65 mg/m³ for the door-machine operator and 0.39 mg/m³ for the quench-car operator. Exposure to CTPV on a battery where there was no coke-side shed was 1.36 mg/m³ for the door-machine operator and 0.19 mg/m³ for the quench-car operator.

Gillies (1983) reported that the mean personal exposures to CTPV for bench workers ranged from 0.7-1.2 mg/m³.

Bjørseth *et al.* (1978) conducted both area and personal sampling for individual polynuclear aromatic compounds (PACs) at a coke plant. The results from area sampling for both particulate and gaseous PACs are presented in Table 5. The presence of 41 PACs was detected in this study. Personal sampling (particulate phase only) in various coke-oven occupations demonstrated the following exposure ranges for total PACs (μg/m³): larry-car operator, 168.20-1044.92; coke-car operator, 4.82-72.83; jamb cleaner, 62.06-243.75; door cleaner, 9.11-17.27; push-car operator, 9.41-62.44; sweeper, 110.92; quench-car operator, 5.75; and wharf man, 362.97. The authors noted that this PAC profile (relative distribution of individual PACs) was quite similar to that found in samples taken from a Söderberg vertical-stud aluminium reduction facility but different from that of an anode-paste plant. The latter difference probably reflects the lower operating temperature at the anode-paste plant.

Bjørseth and Eklund (1979) confirmed the similarities in PAH profiles in air samples from a coke oven and from a Söderberg vertical-stud aluminium reduction facility in a subsequent study, ana-

Table 5. Concentrations of polynuclear aromatic compounds (PACs) (μg/m³) in the atmosphere of a coke plant; stationary sampling at the battery top, 1976[a]

PAC	Range	
	Particulates	Gaseous
Naphthalene	0-4.4	278-1151
2-Methylnaphthalene	0-3.0	43-186
1-Methylnaphthalene	0-0.08	0-86
Biphenyl	0-0.39	0-34
Acenaphthalene	0-30	0-251
Acenaphthene	0-17	6.0-104
Dibenzofuran	0-60	19-81
Fluorene	0-58	23-125
9-Methylfluorene	0-70	0.7-1231
2-Methylfluorene	4.6-20	2.0-10
1-Methylfluorene	0.7-2.0	0-10
Dibenzothiophene	0-76	1.2-49
Phenanthrene	27-891	6.7-276
Anthracene	9.6-314	6.0-91
Carbazole	3.0-19	0-11
2-Methylanthracene	2.8-23	<0.7-16
1-Methylphenanthrene	2.7-21	0-7.0
9-Methylanthracene	3.2-17	0.8-16
Fluoranthene	45-427	0-24
Dihydrobenzo[*a*]- and dihydrobenzo[*b*]fluorene	9.2-115	0-2.8
Pyrene	35-319	0-14
Benzo[*a*]fluorene	9.7-90	0-6.8
Benzo[*b*]fluorene	3.1-61	0-0.3
4-Methylpyrene	0-3	-
1-Methylpyrene	2.5-24	0-0.2
Benzo[*c*]phenanthrene	2.6-49	-
Benz[*a*]anthracene	5.4-156	<0.4-1.6
Chrysene triphenylene	26-189	0-1.8
Benzo[*b*]fluoranthene	5.5-67	0-0.7
Benzo[*j*]- and benzo[*k*]fluoranthene	0-35	0-0.7
Benzo[*e*]pyrene	8-73	0-0.2
Benzo[*a*]pyrene	14-135	0-1.5
Perylene	3.3-19	0-0.1
o-Phenylenepyrene	6.3-74	0-1.2
Benzo[*ghi*]perylene	8.7-45	-
Anthanthrene	2.6-62	-
Dibenzo[*a*,*l*]pyrene	3.6-24	-
Coronene	1.0-19	-

[a]From Bjørseth *et al.* (1978)

lysing the air samples using a glass capillary gas chromatography-mass spectrometer-computer system with a higher separation efficiency than the earlier analytical technique. This study also found a variety of other polynuclear aromatic compounds containing nitrogen, oxygen and sulphur in the atmospheres of the coke plant.

3.2 Other exposures

Formaldehyde, acrolein and aliphatic aldehydes have been detected at various locations throughout a coke plant. The concentrations, which were highest on the topside of the battery, were for formaldehyde, 184 μg/m^3; acrolein, 140 μg/m^3; and the aliphatic aldhehydes, 380 μg/m^3 (Mašek, 1972a). Arsenic, cadmium, mercury and hydrogen cyanide may also occur in low concentrations (Mašek, 1972b, 1974, 1979, 1980). Asbestos, which is used as insulation, may be present on a coke battery (Selikoff & Hammond, 1971).

Amangeldin (1982) measured the concentrations of several contaminants in workplace air during coal charging, door removing and quenching operations. The means of 10-44 samples taken from each operation in a coke plant ranged as follows (mg/m^3): ammonia, 0.33-7.78; naphthalene, 0.48-11.86; carbon monoxide, 7.48-18.20; nitrogen oxides, 1.32-3.60; and phenol, 0.12-0.32.

Mašek and Neiser (1973) found high concentrations of pyridine and its homologues in 18 locations in three coking plants. About 60% of the homologues was composed of 2-picoline. The results ranged as follows (μg/m^3): pyridine, 8-2980; 2-picoline, 35-8256; 3-picoline, 10-837; 4-picoline, 15-1017; 2,6-lutidine, 4-903; 2,5-lutidine, 11-847; 2,4-lutidine, 2,3-lutidine, 2,4,6-kollidine and 2,3,6-kollidine, 15-618.

4. Biological Data Relevant to the Evaluation of Carcinogenic Risk to Humans

The studies of carcinogenicity and mutagenicity reported in this section include only those on samples of complex mixtures taken from the industry.

4.1 Carcinogenicity studies in animals

(*a*) *Skin application*

Mouse: The carcinogenicity of coke-oven emissions (topside and main)[1] was investigated in SENCAR mice and reported in two sequential papers (Nesnow *et al.*, 1982, 1983).

[1]The topside sample represented a mixture of the environmental particle emissions <1.7 μm that occurred during all the coking operations in addition to background ambient air particulate matter. The coke-oven main sample represented a sample of the vapours and other by-products normally recovered in the by-product gas collector during the charging operation (Albert *et al.*, 1983).

The two coke-oven samples were collected from a modern coke oven and included a topside air particulate sample collected by means of two massive air volume samplers placed on top of the coke-oven battery, and a coke-oven main sample collected from a separator located between the gas collector main and primary coolers (Lewtas *et al.*, 1981; Albert *et al.*, 1983). The coke-oven topside sample was extracted with dichloromethane and the extract was then evaporated to dryness under nitrogen; this extract was found to contain benzo[*a*]pyrene at a level of 478 mg/kg. The coke-oven main sample was applied topically to groups of 40 male and 40 female SENCAR mice at 0.1, 0.5, 1.0, 2.0 and 4.0 mg per mouse on a weekly basis, except for the highest dose (4.0 mg), which was applied twice weekly at 2.0 mg per application, for 50-52 weeks (Nesnow *et al.*, 1983). Control groups were treated with acetone (negative control) or benzo[*a*]pyrene (positive control) once weekly for 50-52 weeks. The incidence of skin carcinomas, as shown in Table 6, ranged from 5% to 98%.

Table 6. Skin carcinomas observed following repeated administration to SENCAR mice of samples of coke-oven main for 50-52 weeks[a]

Sample	Dose (mg/mouse per week)	Sex	Carcinomas (% incidence)
Acetone	-	M	0
	-	F	0
Benzo[*a*]pyrene	0.0051	M	93
	0.0051	F	98
Coke-oven main	0.1	M	5
	0.1	F	5
	0.5	M	36
	0.5	F	30
	1	M	48
	1	F	60
	2	M	82
	2	F	78
	4	M	98
	4	F	75

[a]From Nesnow *et al.* (1983)

The tumour-initiating activity of both samples was studied using the two-stage model (Nesnow *et al.*, 1982). Groups of 40 male and 40 female SENCAR mice per treatment group were painted once only with 0.1, 0.5, 1.0, 2.0 or 10 mg of each sample dissolved in 0.2 ml acetone. The 10-mg dose was administered in divided doses of 2.0 mg each for 5 days. Commencing one week later, 2 μg of 12-*O*-tetradecanoylphorbol-13-acetate (TPA) were administered topically twice weekly. Negative controls (treated with TPA only) and positive controls (treated with a single dose of benzo[*a*]pyrene followed by TPA) were included. The incidences of papillomas and of carcinomas were reported at six months and one year, respectively, after treatment with the coke-oven samples. Between 37-40 animals survived per group. Results are shown in Table 7. Both the topside and main samples produced dose-related increases in both papillomas and carcinomas (Nesnow *et al.*, 1982, 1983). For the topside sample, the papilloma incidence ranged from 10-100% and carcinomas ranged from 0-20%, while for the coke-oven main sample, papilloma incidence ranged from 31-100% and carcinomas from 10-65%.

Table 7. Skin tumours observed following administration of a single dose of benzo[*a*]pyrene (B[*a*]P), topside coke-oven extract, and coke-oven main extract to SENCAR mice followed by repeated administrations of 12-*O*-tetradecanoylphorbol-13-acetate (TPA) twice weekly[a]

Sample	Dose mg/mouse	Sex	Mice with papillomas[b] (%)	Mean no. of papillomas per mouse[b]	Mice with carcinomas[c] (%)
TPA alone	0	M	8	0.08	5
	0	F	5	0.05	0
B[*a*]P	0.1	M	95	10.2	30
	0.1	F	97	7.9	25
Topside coke oven	0.1	M	13	0.13	0
	0.1	F	10	0.20	8
	0.5	M	73	1.6	5
	0.5	F	70	1.8	15
	1	M	95	2.6	15
	1	F	72	2.0	3
	2	M	95	4.0	13
	2	F	90	3.5	10
	10	M	100	7.1	13
	10	F	100	7.7	20
Coke-oven main	0.1	M	50	0.63	10
	0.1	F	31	0.38	25
	0.5	M	90	3.7	54
	0.5	F	92	2.2	54
	1	M	87	3.3	53
	1	F	90	3.1	48
	2	M	78	3.1	48
	2	F	100	5.3	45
	10	M	100	8.9	55
	10	F	100	8.1	65

[a]From Nesnow *et al.* (1982, 1983)

[b]Scored six months after initiation

[c]Cumulative score one year after initiation

The sample from the coke-oven main was also tested by repeated weekly skin applications of 0.1, 0.5, 1.0, 2.0 and 4.0 mg for 34 weeks to animals pre-treated (one week earlier) with 50.5 μg benzo[*a*]pyrene. Control groups were treated with a single dose of 50.5 μg benzo[*a*]pyrene alone or followed by twice-weekly applications of 2 μg TPA for 34 weeks. Papilloma incidence was reported to increase from 3% to 100% over the dose range of coke-oven main sample tested. The group receiving a single benzo[*a*]pyrene treatment had no papilloma, and the group receiving benzo[*a*]pyrene treatment followed by TPA had 86% and 97% papillomas (Nesnow *et al.*, 1983).

The water effluent from a coking plant, suspected to contain carcinogens, was passed through an activated carbon filter and the condensate was extracted by benzene and painted twice weekly [amount unspecified] for 22 months on the skin at the nape of the neck of 25 male and 25 female C57 black mice. None of these mice developed skin tumours (Hueper & Payne, 1960).

(*b*) *Inhalation*

Mouse: The carcinogenicity of coke-oven tar (from a US by-product coke oven) was investigated in C3H mice that had been exposed to formaldehyde vapour. This tar sample was selected by the authors because it had been shown to be carcinogenic to mouse skin by topical application in an earlier experiment. [The information available on the skin carcinogenesis experiment of this tar was insufficient for evaluation by the Working Group.] Four groups of mice were used. The first group was an untreated control group, and all mice survived for the 71 weeks' duration of the experiment. A second group of 33 mice was exposed to a coal-tar aerosol administered at a rate of 300 mg/m³ for two hours daily three times weekly in weeks 35-68 of the experiment. Two squamous-cell tumours of the lung, one of which was malignant as well as a pulmonary adenoma occurred in this group. In 10 surviving mice treated until week 71, three squamous-cell tumours of the lung were seen. A third group of 60 mice was exposed to formaldehyde vapour at 100 mg/m³ for one hour daily three times weekly for 35 weeks, at which time 26 mice were still alive. They were then exposed to the coal-tar aerosol at a concentration of 300 mg/m³ for two hours daily three times weekly; one squamous-cell tumour and one pulmonary adenoma were found in the lung of a mouse that died between weeks 47-58. All mice had died by week 68. No lung tumour occurred in mice exposed to formaldehyde at 50 mg/m³ for 35 weeks and then at 150 mg/m³ for up to 68 weeks (Horton *et al.*, 1963).

In another experiment, the tar from the same source as that in the previous paper (Tar 1) and another tar (Tar 2) were separated into phenolic (P-Tar) and non-phenolic (N-Tar) fractions. Both tars were derived from US by-product coke ovens. Aerosols were produced from these fractions and were administered singly or in combination at levels of 120-200 mg/m³ three times weekly for 55 weeks. Six groups of 50 male C3H/HeJ mice were treated with air only or with aerosols of Tar 1, N-Tar 1, N-Tar 1 plus P-Tar 1 re-added, N-Tar 1 plus P-Tar 2 or N-Tar 2 plus P-Tar 1. The experiment was terminated at 55 weeks, but some animals died during the experiment. Attention was given only to lung tumours in animals that died between 46-55 weeks and in those that were killed at termination of the experiment. No lung tumour was observed in the animals exposed to air only. In the 13 mice exposed to Tar 1 that survived beyond 46 weeks, 15 tumours were found; 12 of these were adenomas and three were adenocarcinomas. In addition, five animals with squamous metaplasias in the trachea were also found. In the 20 mice that were exposed to N-Tar 1 and had survived beyond 46 weeks, 16 pulmonary adenomas and two squamous metaplasias were found. In the 19 surviving mice that were exposed to N-Tar 1 plus P-Tar 1 re-added, 14 had pulmonary adenomas, one a pulmonary adenocarcinoma and five, squamous metaplasias. In the 25 surviving mice exposed to N-Tar 1 plus P-Tar 2, 14 had pulmonary adenomas, one had a pulmonary adenocarcinoma and seven had squamous metaplasias. In the 23 mice exposed to N-Tar 2 plus P-Tar 1, 14 had pulmonary adenomas and four had squamous metaplasias (Tye & Stemmer, 1967).

Groups of 75 CAF_1-JAX and an equal number of ICR-CF_1 female mice were exposed continuously for 90 days to concentrations of coal-tar aerosol of 0.2, 2.0 and 10 mg/m³, and were observed for a further 21 weeks. Controls of each strain of mice were available. The coal-tar used to generate the aerosol consisted of a composite mixture collected from the effluents of several different coke ovens in the area of Pittsburgh, PA, USA. The effluents were trapped in air-collection devices using a chilled water spray. After settling and separation of the liquid phase, the various coal-tar samples were blended together with 20% by volume of the benzene, toluene, xylene (BTA) fraction of the coke-oven distillate. The aerosol was generated from this blend; a minimum of 99% of the droplets were 5 μm or less in size. Data are available only with regard to skin tumour response: the numbers observed were 1/61, 14/75 and 44/55, respectively, in the treated ICR-CF_1 mice, and 3/225 for the controls. For CAF_1-JAX mice, the numbers of skin tumours were 0/75, 3/65 and 18/43 in the exposed groups, respectively, and 0/225 in the controls (MacEwen *et al.*, 1976). [No solvent control was reported.]

A group of 75 female ICR-CF_1 mice and a group of 50 female CAF_1-JAX mice were exposed to the coal-tar aerosol described in the previous study at a concentration of 10 mg/m^3 for six hours per day, five days per week for a total of 18 months; 75 ICR-CF_1 and 50 CAF_1-JAX controls were available. Alveologenic carcinomas were produced in 26/61 ICR-CF_1 mice, and in 27/50 CAF_1-JAX mice examined. The numbers of these tumours in the ICR-CF_1 and the CAF_1-JAX control mice were 3/68 and 8/48, respectively. The exposed and control groups did not differ in the incidence of other types of tumours. Skin tumours were produced in 5/61 treated ICR-CF_1 mice and 2/50 treated CAF_1-JAX mice examined and in 3/68 and 1/48 respective controls. All tumours were examined microscopically (MacEwen *et al.*, 1976).

Rat: A group of 40 male and a group of 40 female CFE strain Sprague-Dawley weanling rats were exposed to the same aerosol as the mice at a concentration of 10 mg/m^3 for six hours each day, five days per week, for 18 months. An equal number of male and female control rats was available. Among treated rats, all 38 males and 31 of 38 females examined histologically developed squamous-cell carcinomas of the lung. In addition, eight unspecified tumours were reported in the male rats and three mammary fibroadenomas and two unspecified tumours in the female rats. In the control group, no tumour was found in 36 males examined; five tumours (one sebaceous-cell carcinoma, one intra-abdominal carcinoma, one mammary fibroadenoma, one mammary adenocarcinoma and one other tumour) occurred in 37 females examined (MacEwen *et al.*, 1976).

4.2 Other relevant biological data

(*a*) *Experimental sytems*

No data were available to the Working Group on toxic or reproductive effects.

Tests for genetic and related effects in experimental systems

Two coke-oven samples collected both on the topside of a modern coke-oven battery and from the gas-collector main (Lewtas *et al.*, 1981; Albert *et al.*, 1983), already referred to in section 4.1(*a*), were also studied in a series of short-term genetic and cell transformation tests, as described below.

The topside coke-oven sample was mutagenic to *Salmonella typhimurium* strains TA98 and TA100 both with and without an exogenous metabolic system (Claxton, 1981). A series of other topside coke-oven samples taken from the same coke battery were also reported to be mutagenic to strain TA98 with and without metabolic activation. The mutagenicity of these samples was not affected by the electric charge on the electrostatic precipitation in the air sampler (Jungers *et al.*, 1981). Another coke-oven main sample was also mutagenic to strain TA98 both with and without the metabolic system; both coke-oven main and topside samples demonstrated increased mutagenicity in the presence of metabolic activation in strain TA98 (Albert *et al.*, 1983). Evaluation of the mutagenicity of fractions from the coke-oven main sample showed that the basic and polar (cyclohexane insoluble) fractions were significantly more mutagenic in the presence of the metabolic system as compared to other fractions (including the fraction containing unsubstituted polynuclear aromatic hydrocarbons (Austin *et al.*, 1984).

The topside coke-oven sample did not induce mitotic recombination in *Saccharomyces cerevisiae* D3 (concentrations tested, 100-1000 μg) with or without metabolic activation (Mitchell *et al.*, 1980, 1981).

The topside coke-oven sample (125 and 250 μg/ml) induced DNA strand breakage in primary Syrian hamster embryo (SHE) cell cultures, as determined by sedimentation in alkaline sucrose gradients (Casto *et al.*, 1980, 1981).

Both topside and main coke-oven samples tested at five to seven concentrations (1-160 μg/ml) were mutagenic to L5178Y mouse lymphoma cells ($TK^{+/-}$); the activity was significantly increased in the presence of an exogenous metabolic system (Mitchell *et al.*, 1980, 1981; Albert *et al.*, 1983). The topside coke-oven sample also induced mutations to 6-thioguanine resistance in Chinese hamster ovary cells (50-200 μg/ml) (Casto *et al.*, 1980, 1981), and to ouabain resistance in BALB/c 3T3 cells (10-1000 μg/ml) (Curren *et al.*, 1980, 1981).

Both topside and main coke-oven samples induced sister chromatid exchange in Chinese hamster ovary cells with and without metabolic activation (Mitchell *et al.*, 1980, 1981; Albert *et al.*, 1983).

The topside coke-oven sample caused enhancement of viral transformation of SHE cells (31.2-500 μg/ml) (Casto *et al.*, 1980, 1981) and induced morphological transformation of BALB/c 3T3 cells (Curren *et al.*, 1980, 1981).

(b) Effects in humans (other than cancer)

Toxic effects

Walker *et al.*, (1971) studied 881 men in two coking plants in the UK. Among 169 men, 148 of whom were smokers, there was evidence, by questionnaire and lung function test, of bronchitis; 80 of 334 oven men had bronchitis, compared with 89 of 547 other coke workers. Cigarette smoking alone and in combination with dusty work appear to be causative factors for bronchitis in coke-production workers.

Nilsen *et al.* (1984) found a significantly increased number of macrophages in the expectorates of 43 exposed coke-oven workers compared with 36 controls [occupations not specified]. A more than additive effect of smoking and exposure was observed.

Mutagenicity in urine

Urine samples from 12 non-smoking coke-plant workers who worked on or along the coke batteries were found to have significantly higher mutagenic activity in *Salmonella typhimurium* strain TA1538 in the presence of a liver metabolic system than did samples from a control group of 35 office and laboratory workers who were non-smokers. No significant difference was found in studies using strain TA100 (Kriebel *et al.*, 1983).

XAD-2 resin extracts of urine samples collected from coke-plant workers (seven battery workers, ten truck drivers and three shift foremen) both before and after work were tested for mutagenic activity in *S. typhimurium* strain TA98 in the presence of an Aroclor-induced rat-liver metabolic system. The mutagenic activity of extracts from exposed workers did not differ significantly from that of extracts from non-exposed workers [jobs not specified]. The mutagenic activity of urine extracts from smokers was significantly higher than that from non-smokers (Møller & Dybing, 1980). [The Working Group noted that only seven coke-oven battery workers may have been heavily exposed.]

4.3 Case reports and epidemiological studies of carcinogenicity in humans

(*a*) *Epidemiological studies on the destructive distillation of coal*

For over 200 years, skin cancers have been recognized to be associated with occupational exposure to tars and pitches related to the destructive distillation of coal. Studies from the early twentieth century drew attention to additional cancer sites, but they often did not distinguish the specific industry from which cases originated. It therefore seemed prudent to consider the early studies together.

Early authors alluded to case reports of scrotal cancer in previous work. O'Donovan (1920) reviewed the findings of Butlin (1892) and indicated that 34 of the 39 scrotal cancer cases admitted to St Bartholomew's Hospital, London, UK, were of industrial origin. The author described 16 additional skin cancers from exposure to tars and pitches. After the enactment in the UK of the Workmen's Compensation Act of 1907, 46 cases of scrotal cancer and 'epitheliomatous ulceration' were reported in 1907-1911 and 64 cases in 1914-1918. Of the latter cases, 44 were from two factories (O'Donovan, 1928). The later paper described the cases from the combined studies and found that, of the 33 cases for which a job was identified, nine were in tar workers.

Henry (1947) analysed 3753 cases of skin tumours that were reported to the Factory Inspectorate in the UK between 1920-1945. He attributed 2229 cases (59.4%) to tar and/or pitch exposure. The industries included tar distillery, patent fuel, as well as gas-works, coke ovens and coal carbonization. Scrotal lesions represented 21.3% of sites reported. The author noted that the time from first exposure to the time of diagnosis of a skin tumour was shorter for the tar and pitch products (20 years) than for oil exposures (50 years). [Compensation practices would have influenced the representation of various jobs and industries among the cases.]

Ross (1948) noted that in the UK the reported number of compensation cases of 'epitheliomatous ulcerations' rose during 1920-1945 from 89 in the first five-year period to 449 in the last five years. The deaths, however, which initially were 15, had risen to 62 in 1931-1935 and fallen to nine in 1941-1945. The author pointed out that it is difficult to know whether these changes were due to variation in reporting practices or to a true increase in disease. Ross also reported the frequency of lesions by three-year periods from 1936-1947 in an average of 170 personnel employed in the manufacture of carbon products. The frequency of epithelioma ranged from 1.2-4.7% and that of papilloma from 10.6-23.5% with total numbers over the whole period of 16 and 102, respectively.

Henry *et al.* (1931) collected death certificates relating to 13 965 bladder and prostate cancer deaths in England and Wales for the period 1921-1928 and compared the observed distribution of occupations recorded on these certificates to those expected on the basis of the occupational groupings in the 1921 census. Although 11 429 male deaths from either cancer were included, the numbers of deaths among many of the individual job categories were few. [The consistency of the finding of elevated standardized mortality ratios (SMRs) for bladder cancer in nine of eleven occupational groups with exposure to coal gas, tars and pitches is noteworthy.] Patent-fuel labourers, tar-distillery workers and most job classifications from the gas-works had SMRs between 117 and 400. Only chimney sweeps and gas-works inspectors had ratios below 100. The SMR for bladder cancer in coke-oven workers was 122. The combined SMR can be calculated from the data provided by the authors: for these selected occupations with exposures to tars and pitches, the SMR is 144 (63/43.8), which is significant. There was no apparent association with a risk of prostate cancer in the same occupational groups in which a risk of bladder cancer occurred. A special study was made of 50 cases of bladder cancer with enquiry made of the work histories by interviews with work mates, spouses and employers; 44 were traced, and in 41 the occupational categorization was considered to be accurate.

Kennaway and Kennaway (1936, 1947) reported two studies using a similar method of analysis to that used by Henry *et al.* (1931), but selecting deaths from lung and laryngeal cancer in England and Wales and examining the distribution of occupations from the 1931 census. The deaths among males aged 20 years or more included 23 549 lung cancers and 14 869 laryngeal cancers in the total period 1921-1938. The number of lung cancer cases increased by 58% between the first 12-year period and the last six-year period, which would represent a real increase of 116%. The number of deaths from laryngeal cancer decreased slightly during this period. In both papers, the authors examined the developing epidemic of lung cancer in that period, in an attempt to relate that event to occupation. The investigation identified several occupations with exposure to tars and pitches which were associated with high mortality ratios (SMRs). The lung cancer SMRs were high for patent-fuel workers (571), gas stokers and coke-oven chargers (284), and several other occupations associated with gas or gas production, such as gas-works engine drivers, gas-works foremen, gas producers and gas fitters (138-202). The category of 'tar distillery and coke-oven workers' had a very low SMR (52), despite a risk of bladder cancer common to several of these groups. SMRs for laryngeal cancer were not high in the same occupational groups. Gas stokers and coke-oven chargers who had high SMRs for lung cancer also had high ratios for laryngeal cancer (213), but, among the occupational groups exposed in the gas-works, only engine drivers had a SMR for laryngeal cancer above 110 (227).

[The studies presented in the three papers above have common problems: even with deaths accumulated over an 18-year period, single categories of deaths were often so small that few deaths could have occurred from a specific cancer. Also, comparison is made between occupation stated on a death certificate and census data. The accuracy of diagnosis on a death certificate could be questioned, and no information is available on length or type of occupational exposure.]

(*b*) *Epidemiological studies of coke-oven workers*

Kuroda (1937) reported the first epidemiological study of lung cancer related to employment in a gas-generator plant of a coking operation of a steel company in Japan. His cross-sectional study of current workers identified 61 malignancies in the period 1931-1935, of which 12 occurred in the lung. The most remarkable observation was that all lung cancers occurred in workers in the gas-generator works. A crude distribution of cancers by site for this section of the plant compared to the rest of the plant indicates that 80% (12/15) were lung cancer, 13% (2/15) stomach cancer, and 7% (1/15) liver cancer. Comparable figures for the rest of the plant were no lung cancer, 65% (30/46) stomach cancer, 24% (11/46) liver cancer and 11% (5/46) other cancers. Workers with lung cancer had been exposed in the gas-generating plant for an average of 15.6 years. [Statements in the table of morbidity rates are unclear with regard to the total number of workers in whom these cancers appeared. Five-year incidence in the 22 867 workers was calculated to be 0.53 per 1000 for lung cancer and 2.67 per 1000 for all cancers.]

Kawai *et al.* (1961) described an additional six lung cancer cases that occurred in 1946-1953 in workers in the same gas-works and four more cases in 1954-1960 after the plant was closed in 1953. These cases, added to the 21 originally reported in 1933-1937 (Kawahata, 1938), resulted in a total of 31 lung cancer cases among the workers from the gas-works, with no reported cases during the eight-year war period. A subsequent paper by Kawai *et al.* (1967) followed through 1965 the cohort of 504 men who were working in the gas-generating plant at the time of its closure in 1953. Deaths were identified for all workers through plant records until the age of retirement at 55 years. The cancer rates were compared with those of the remaining 25 760 workers in the steel plant. For men over the age of 46 years, the risk of lung cancer was 44 times higher than that expected on the basis of the rates in the remaining plant workers. No case occurred in men under the age of 45 years. The risk increased with length of time worked in the gas-generator plant, with a risk ratio of 18 for those with employment of less than 10 years, 32 for those with 10-19 years and 136 for those with 20 years or more. There was a significant increase in risk of lung cancer deaths between those in the two longest employment classes, 10-19 years and 20 years or more. The cancers for which histological information was available were few and did not reveal major differences in cell type between the lung cancers in former gas-works employees and those of other workers. Cigarette smoking was not controlled for in this study but had not affected the risk estimates in the previous analysis of Kawai *et al.* (1961). There was no excess of cancer of the stomach, prostate or liver. [The risks reported in this study may underestimate the true risk to those exposed because no follow-up was done beyond the age of 55 years, the age of retirement, despite the fact that observed risk increased with increasing age. The authors also indicated that no risk was observed for workers under the age of 45 years; since the expected number of deaths was very low (0.04), it is impossible to be sure that there was not an undetected risk in this young age group.]

Reid and Buck (1956) published a report of cancers based on deaths during 1949-1954 among current and retired workers at coking plants in England. A population of 8000 current workers, for whom information on age and current job was available, was identified from a plant census in 1952. Complete job histories were obtained on a 10% sample. A total of 337 deaths occurred among current workers. Information was available on the deaths of 214 retired workers. For workers still employed in the study period, standardized mortality ratios (SMRs) were determined using a 'large industrial population' for comparison. The SMR for lung cancers for those presently working in a coke-oven area was 80 and for all cancers was 150. For those who had 'ever' worked in coke ovens, based on an estimate from a 10% sample, the SMR was 140 for lung cancer in coke-oven workers and 125 for all cancers. The proportion of lung cancers among all cancer deaths for retired workers was 30% (15/51) compared to 25% (3/12) for coke-oven workers. The authors concluded that there was no increase in risk of cancer associated with work at the coke ovens. [The design of this study would seriously underestimate risks, since the SMRs refer to workers active in 1949 who then had only six years of follow-up during which death might have occurred. The proportional death rates in retired workers are difficult to evaluate because they are not age adjusted. The eligibility for retirement could influence the distribution of causes of death among retired workers.]

Lloyd and Ciocco (1969), in the first of a series of papers on a large occupational study of the steel industry in the Pittsburgh area (Allegheny County, PA, USA), described a historical cohort study of 59 072 employees of seven plants in three steel companies. The population study was initiated in 1953 and comprised all current employees at that time, all of whom were followed, with 99.8% ascertainment of vital status, through 1961. Using 53 categories, Lloyd *et al.* (1970) classified each individual in the population of steel workers by work area at the time of the start of follow-up in 1953 and according to those areas in which he was employed for a minimum of five years between start of employment and the end of follow-up in 1961. For work areas in which the SMR for all causes of mortality was significantly higher than 100, further SMRs were calculated for broad categories of disease, including malignant neoplasms. For males who had worked in the coke plant for five years or more, the SMR for all causes was 99 for whites and 122 ($p < 0.05$) for non-whites, the SMR for malignant neoplasms was

102 for whites and 204 ($p < 0.01$) for non-whites. All the cancer excess in non-whites was due to malignant neoplasms of the respiratory system (SMR, 342; 25 observed).

A more detailed review of the risk in the coke plant by Lloyd (1971) indicated that a total of 3530 employees had worked in that area, 2048 at the ovens and 1482 in non-oven areas, employed in 1953 or prior years. The SMR for respiratory cancers among coke-oven workers was 248 (33 deaths) compared to an SMR of 47 (4 deaths) in non-oven workers employed in the coke plant. Most of the risk occurred in non-white workers (SMR, 298 in non-white compared to 160 in white coke-oven workers). Jobs were reclassified on the basis of exposure to the ovens as well as the location of the job in the working environment in relation to effluents from the oven. Data were reported for workers who had worked for five years or more in the coke plant. If the workers had worked five years or more full-time on the top of the oven ('topside', the site of coal loading), the highest SMR was for lung cancer (1000, based on 15 deaths). There was a significant excess of intestinal cancer excluding rectum (SMR, 316) among coke-plant workers who had never worked at the ovens. The numbers of cancers at any other site among those employees who worked topside were too few to produce meaningful results. Two kidney cancers occurred in those who had worked at any time topside on the ovens, compared to an expected number of 0.6. Two lymphomas occurred in the same group, compared to 0.4 expected. The author has suggested a possible relationship between the excess risk of lung cancer in tar-exposed populations and the temperature of the process at which the tar is produced.

Redmond *et al.* (1972) expanded the initial study of steelworkers and included for analysis not only coke-oven workers at two of the original plants of Lloyd but also 10 additional plants from various parts of the USA or Canada. The coke-oven workers in the added population had worked between 1951-1955. For the two original plants, all non-oven workers were used as controls. For the additional plants, two non-oven workers of the same sex, race and calendar time of hire were selected as controls for each oven worker in this matched design. The total populations of coke-oven workers (4661) and non-exposed workers (25 011) were followed through 1966 - about 13 years. [Unlike those in previous analyses of studies of steelworkers, the relative risks in this analysis were weighted by a 'precision' factor for each plant, a procedure which increased the SMR over that calculated directly.] The population in this study had a significant excess relative risk for lung cancer of 2.85 (69 observed, 41.5 expected) with excess risks for both white (2.06; 17/10.8) and non-white males (3.35; 52/30.7). Genito-urinary cancers also occurred at a significantly high relative risk of 2.05 (21/13.3). For non-white workers who had worked topside of the ovens throughout their work history, the relative risk for lung cancer was 7.68, which is significant. For white workers, the relative risk was 4.49, based on only three deaths. There was a gradient of response, which might be expected on the basis of exposure, with men who had worked full-time topside having a relative risk of 7.24 (35 observed/10.4 expected), topside at some time, 2.14 (7/3.7) and those who had always worked at side ovens, 1.73 (27/19.4). The relative risk for those with employment of five years or more on the topside was 6.87 (25/7.4). The relative risk of kidney cancer was 7.49 (8/2.6; significant) in total oven workers. The numbers were too small for further subdivision, but all the cases appeared in coke-oven workers, regardless of position in relation to the ovens. There was no elevated risk of bladder or skin cancer. There was a non-significant excess risk of prostatic cancer in those who had worked five years or more in the coke oven: relative risk of 1.83 (11/7.4). The only excess of digestive-system cancers (relative risk not statistically significant, 1.37) that was observed in this population occurred among side-oven employees who had worked five years or more but had never worked topside.

Redmond *et al.* (1976) examined workers at the seven steel plants included in the Lloyd study, who were exposed in the by-product coke plant, subdivided into coke-oven and non-oven (coal handling, by-products and other non-oven) areas according to their job assignment. A statistically significant five-fold increase in deaths from kidney cancer was found for the combined population of coke-oven and non-oven workers compared with controls. There were significantly elevated relative risks for non-oven workers for buccal and pharyngeal cancers (3.87). For employees who had worked for five years or more in the coke plant, the non-oven workers had a significant excess of cancers of the digestive organs (1.62), and oven workers had a non-significant excess of prostate cancer (1.42). The excess of kidney cancer was statistically significant for all the coke-plant workers (4.50; 6/1.6); four deaths were seen among coke-oven workers, and two among non-oven workers. The excess in digestive cancers in non-oven workers was attributed to high relative risks for large intestinal (2.93) and pancreatic cancer (4.55), both of which were significant. The numbers were too small to determine whether these risks occurred in by-products, coal handling or other non-oven jobs. The intestinal cancer risk was significantly higher for foreign-born, total non-oven white workers (4.40); that for the native-born workers was also raised.

In an attempt to relate specific agents to the lung cancer mortality of coke-oven workers, Mazumdar *et al.* (1975) tested the levels of benzene-soluble particulates in ten plants included in the expanded population of coke-oven workers described by Redmond *et al.* (1972). The particulates were measured for each job by exposure area. These levels by job were then totalled for the three exposure areas - topside oven, side ovens and other locations - and then averaged. An individual was assigned this average dose cumulated over the months worked. The averages were significantly different for different work sites; topside, 3.15 mg/m^3; sidemen who worked near the ovens, 1.99 mg/m^3; and sidemen who worked away from the ovens, 0.88 mg/m^3 (see also pp 00 of this monograph). There was no association between mortality from lung cancer and dose in white workers, but there were only 62 workers in the highest exposure category. For the non-whites, the relative risk of lung cancer between those with the highest and lowest cumulated dose level was 13.65 (54.6/1000 to 4.0/1000). No increased risk of lung cancer was seen for individuals with cumulative doses below or equal to 199 mg/m^3-months.

Lloyd (1980) presented results from the updated study of the coke-oven population of his earlier study (Lloyd, 1971) through 1975. With 30.9% of the original population from Allegheny County plants deceased, the relative risk for lung cancer in men employed for five years or more in 1953 and working fulltime topside was 6.94, which was significant and based on 25 deaths instead of 15 in 1961. [The accumulation of additional cases appears to be low, since 15 cases were identified in the first nine years of study and 10 cases in the last 14 years, as the population continued to age. This may suggest a decreasing risk in more recent periods.] Lloyd also pointed out the possible underestimation of relative risk that could result from the use of internal controls, since a high risk of lung cancer has been demonstrated in many other work areas in steel mills compared to rates in the general population.

Detailed results from the most recent follow-up of the workers studied by Redmond and her colleagues, extending the observation period for mortality through 1975, have been reported (Redmond *et al.*, 1981). Findings of significant excess risks of cancers of the lung and of the genito-urinary organs, especially kidney, among coke-oven workers were consistent with earlier reports. For the two Allegheny County coke plants (the largest two of the 12 coke plants), the excess risks for workers employed at the coke ovens for five or more years for lung cancer and kidney cancer were 2.63 (63/28.3) and 3.55 (6/1.8), respectively. Excesses of deaths from cancers of the large intestine and pancreas were seen in white workers only in coal-handling and all other non-oven areas, the relative risks being 2.11 ($p < 0.05$) and 2.44 ($p < 0.05$). Redmond *et al.* (1976) had previously reported excess risks at these sites among white by-product workers.

Radford (1976) studied the distribution of deaths among active and retired union employees of a large US steel company not included in the Lloyd study. Certificates for 74% of the known deaths in one year (1973-1974) were examined, making a total of 649 deaths. The data were analysed by proportional mortality ratios, using the distribution of specific cancer deaths from a nearby large metropolitan area for 1969-1972. The proportions of specific cancers in the 'other cancer' category of deaths were estimated for 1973 from previous US rates and time trends. When the age-adjusted proportional mortality ratios (PMRs) were calculated for common cancers but not for rare sites, they were found to be elevated for lung (149), oesophageal (169) and other cancers (220) for both races and for leukaemia in blacks (179). Lung cancer and 'other cancers' were the only groups that included more than 10 deaths. Statistically significant crude PMRs in white male workers in the basic steel plant were elevated for bladder and kidney cancers, being 5.3 and 2.5 times the municipal ratios. No detailed information was given on worksites for these cancer deaths. [PMRs frequently misrepresent the true risk of death by cause in a population. The situation could be aggravated, as in this case, by under-identification of deaths, loss of terminated workers, and lack of age corrections. It is difficult to assess how the reported excesses relate to coke-oven workers.]

A report by Sakabe *et al.* (1975) included a review of cancer cases among retired coke-oven workers from 11 companies, four iron and steel works, four city gas companies and three coke manufacturing chemical companies in Japan. The historical cohort study included 2201 retired workers who had left the companies between 1947-1973. The SMR for lung cancer was 128 (based on 15 cases among 2178 retired workers between 1949-1973). Work in coke ovens in the iron and steel industry carried a significantly elevated relative risk (2.37), but there was no excess risk of lung cancer in gas-company coke-oven workers. Ten of the 12 patients for whom a history was available were smokers. The report included a study from the iron and steel industry in which retired workers in a steel plant had 1.9 times the risk of lung cancer compared to the general population (6 observed, 3.16 expected).

Davies (1977) studied the coke-oven workers in two steel plants in South Wales. The historical cohort study included all workers from 1954 through 1965 and compared mortality experience to that of England and Wales. The total population of coke-oven workers was 610. The author reported no excess of lung cancer but an elevated SMR for colorectal cancer (192; five deaths) and bladder and kidney cancers (252; three deaths). [The numbers were small in this study, and there was a significantly reduced level of cardiovascular disease, which may represent a selection factor. In addition, the plants differed in overall mortality, which could suggest a problem in death ascertainment. The number of person-years is higher than would be expected on the basis of the size of the population and the number of total years of follow-up, suggesting that this study might require further evaluation.]

Axelson *et al.* (1979) studied 263 men who were employed at a Swedish coke-oven plant for at least three months during 1960 or later. The observation period was 1961-1976 with a 98.1% follow-up. A total of 20 tumours were noted, with three in the lung. There was one case of lung cancer among those with high exposure. The two cases of lung cancer in the sub-cohort with medium exposure represent a significant excess. [No information on smoking history was available.]

Hurley *et al.* (1983) reported on the mortality of 2842 (2753 traced) male coke-oven workers employed continuously at 14 steel plants in England, Scotland and Wales between 1 January 1966 and 31 July 1967, who had been employed for 19 months or more at the time of enrolment. A second population of 3925 (3855 traced) men employed on 1 January 1967 at 13 coking works in England and Wales was chosen. The workers in the two populations have been followed for 12 and 13 years, respectively. The SMR for lung cancer for all coke-oven workers was 117 (167 deaths) using the total population of Scotland and of regions of England and Wales for comparison and 105 if male semi-skilled labourers in the 1970-1972 census were used for comparison. A comparison between oven workers, by job, using the total group as comparison resulted in a higher ratio for part-time oven than for full-time oven workers, but the difference is not statistically significant. The ratios for lung cancer were low for those employed less than five years in ovens and high for all after five to nine years, with no sign of increasing ratios with increasing time worked after five years in the steel plants. An excess of bladder cancer was found among the coke workers (SMR, 161; 12 deaths), as well as of cancer of the buccal cavity and pharynx (SMR, 144), but the numbers were small. An excess of stomach cancer was seen among the coke workers in the non-oven group (SMR, 136; 19 cases). In the coke group, maintenance workers were removed from those included in part-time oven work, but the results did not change. Limited data on smoking histories were available but did not account for the differences. Variability in ratios by plant was marked. [The Working Group noted the short duration of follow-up.]

A general population study of lung cancer was reported by Menck and Henderson (1976) for Los Angeles County, USA, for the period 1968-1973. Deceased cases were included for the period 1968-1970, and incident cases identified from the cancer registry were included for 1972-1973. Mortality and incidence ratios were calculated and combined into a single value (SMR) for each occupation and each industry. The expected distribution of occupation and industry in the population was estimated from a sample of 31 216 white males aged 20-64 years from the 1970 census. The jobs of cases were determined from death certificates or hospital records. There were 24 occupations and 13 industrial occupations associated with significantly high SMRs; several were thought to have involved exposure to polynuclear aromatic compounds: the SMRs were 496 for roofers, 279 for workers in the steel industry and 160 for those working with petroleum and coal products. Despite the large number of cases (3938), occupation was not recorded for 17.5% and industry was not given for 31.0%. [Subdivision into occupations or industries does not provide sufficient numbers or precision in classification of jobs to characterize many exposures in this type of study. The sources of information on occupation differ for the numerator and denominator of the ratio.]

A recent community case-control study in Eastern Pennsylvania, USA (Blot *et al.*, 1983) reported the job and personal histories of 360 cases of lung cancer and of controls ascertained from death certificates in 1974-1977 (the analyses were based on 335 cases and 332 controls). The geographical location for study was selected because of the risk of employment in the steel industry. The matched-pair design controlled for age, race, sex, county of residence and year of death. The authors were able to adjust for confounding variables such as smoking, education and employment in multiple industries, but evaluation of the risks within a single industry was limited with this design. The risk of lung cancer from employment in the steel industry was 2.2 times higher than expected on the basis of numbers in controls, and this was significant. The risks were significantly high for those employed in the industry before 1935 (2.7) and for 20 years or more (2.2). The excess did not disappear when corrected for smoking. When cases were classified by job longest held, there was no excess risk for coke-oven workers (1.2) but there was for foundry (7.1) and steel-plant furnace workers (2.6). When these ratios were adjusted for all variables, using linear logistic analyses, there was a relative risk of 1.9 for usual employment in steel as compared to all other jobs. The risk was common to all cell types of lung cancer but was highest in association with small (oat)-cell cancers. In the selected study area, longest employment in the steel and zinc industry accounted for 28% of the cases. [Data on work history obtained from next-of-kin, as in this study, may result in imprecise work histories. The coke-oven risk in other US steel plants was associated with exposure of non-whites, since so few whites worked at the ovens in other steel studies in the USA. This population did not include non-whites and so the coke-oven risk may not be apparent for that reason.]

5. Summary of Data Reported and Evaluation

5.1 Exposures

The vast majority of coke is produced in by-product (slot) coke ovens. Substantial airborne exposures to polynuclear aromatic hydrocarbons have been measured in occupations in the production of coke. The highest exposures have been reported for workers on the topside of the coke-oven battery. The presence of other substances (including aromatic heteronuclear and substituted aromatic compounds) in the workplace air has been reported.

5.2 Experimental data

A sample from a coke-oven main was tested for carcinogenicity in mice by topical application and produced skin carcinomas. Samples obtained either from the topside or main of a coke oven showed initiating activity in a mouse-skin two-stage model. Samples of tar from coke ovens were also tested by inhalation in mice and rats, producing benign and malignant tumours of the lung; in one of these inhalation studies in mice, skin tumours were produced.

A sample from the topside of a coke oven was mutagenic to *Salmonella typhimurium* and in several mammalian cell systems (L5178Y mouse lymphoma, Chinese hamster ovary and BALB/c 3T3); it also induced DNA damage in Syrian hamster embryo cells and induced sister chromatid exchange in Chinese hamster ovary cells. The sample caused morphological transformation in BALB/c 3T3 cells and enhanced viral transformation of Syrian hamster

Overall assessment of data from short-term tests on a topside sample from a coke oven and a sample from a gas-collector main[a]

	Genetic activity			Cell transformation
	DNA damage	Mutation	Chromosomal effects	
Prokaryotes		+		
Fungi/ Green plants				
Insects				
Mammalian cells (*in vitro*)	+[b]	+	+	+[b]
Mammals (*in vivo*)				
Humans (*in vivo*)				
Degree of evidence in short-term tests for genetic activity: *Sufficient* for both samples				Cell transformation: Positive

[a]The groups into which the table is divided and '+', '–' and '?' are defined on pp. 16-17 of the Preamble; the degrees of evidence are defined on pp. 17-18.

[b]Coke-oven topside sample only

embryo cells. A coke-oven sample taken from the gas-collector main was also mutagenic to *Salmonella typhimurium* and in mammalian cells (L5178Y mouse lymphoma) and induced sister chromatid exchange (in Chinese hamster ovary cells).

5.3 Human data

Chronic bronchitis occurs among coke-production workers, particularly in those who smoke.

Case reports of tumours of the skin (including the scrotum), bladder and respiratory tract in association with employment in industries involving the destructive distillation of coal suggested a link between that industry and human cancer. Despite their methodological shortcomings, the descriptive epidemiological studies based on death certificates corroborated these early suggestions.

The site at which excess cancer rates have been identified most commonly among workers in coke production is the lung. All but two of the relevant analytical epidemiological cohort studies provide evidence that work in coke production carries a significantly elevated risk of lung cancer. The two studies showing no excess suffered from serious methodological limitations. The risk for workers in the coke-oven area varied from three- to seven-fold, the highest risk being for men employed for five years or more and working fulltime on the topside of the coke oven. Few of the studies corrected for smoking.

Excess risk for kidney cancer has been associated with work in coke plants. In one study, a seven-fold increase in risk was seen for all workers employed for five years or more at coke ovens.

In single studies, excess risks were reported for cancers of the large intestine and pancreas.

5.4 Evaluation[1]

There is *sufficient evidence* from observations made in the first half of this century that exposures to coal-tar from the destructive distillation of coal give rise to skin cancer in humans.

The available epidemiological studies provide *sufficient evidence* that certain exposures in the coke-production industry are carcinogenic to humans, giving rise to lung cancer. A possible causative agent is coal-tar fume. There is *limited evidence* that such occupational exposures produce cancer of the kidney, and *inadequate evidence* that they result in intestinal and pancreatic cancers.

There is *sufficient evidence* that samples of tars taken from coke ovens are carcinogenic to experimental animals, producing lung and skin tumours.

A number of individual polynuclear aromatic compounds for which there is *sufficient evidence* of carcinogenicity in experimental animals have been measured at high levels in air samples taken from certain areas in coke production plants.[2]

The available evidence indicates that certain exposures in the coke production industry are carcinogenic to humans.

---

[1]For definitions of the italicized terms, see the Preamble, pp. 15 and 19.

[2]See the Appendix, Table 2.

6. References

Albert, R.E., Lewtas, J., Nesnow, S., Thorslund, T.W. & Anderson, E. (1983) Comparative potency method for cancer risk analysis. *Risk Anal.*, *3*, 101-117

Amangeldin, S.K. (1982) Physiologic and hygienic characterization of the working conditions of operators of modern coke ovens (Russ.). *Gig. Tr. prof. Zabol.*, *3*, 13-17

Austin, A., Claxton, L. & Lewtas, J. (1984) Mutagenicity of fractionated organic emissions from diesel, cigarette smoke condensate, coke oven and roofing tar in the Ames assay. *Environ. Mutagenesis* (in press)

Axelson, O., De Verdier, A., Sundell L. & Tallgren, U. (1979) The mortality pattern among employees in a Swedish coke-oven works (Sw.). *Nord. föret. hälsov.*, *2*, 5-12

Bjørseth, A. & Eklund, G. (1979) Analysis for polynuclear aromatic hydrocarbons in working atmospheres by computerized gas chromatography-mass spectrometry. *Anal. chim. Acta*, *105*, 119-128

Bjørseth, A., Bjørseth, O. & Fjeldsted, P.E. (1978) Polycyclic aromatic hydrocarbons in the work atmosphere. II. Determination in a coke plant. *Scand. J. Work Environ. Health*, *4*, 224-236

Blome, H. (1981) Measurements of polycyclic aromatic hydrocarbons at places of work - Assessment of results (Ger.). *Staub-Reinhalt Luft*, *41*, 225-229

Blot, W.J., Brown, L.M., Pottern, L.M., Stone, B.J. & Fraumeni, J.F., Jr (1983) Lung cancer among long-term steel workers. *Am. J. Epidemiol.*, *117*, 706-716

Braszczyńska, Z., Lindscheid, D. & Osińska, R. (1978) Evaluation of the influence of new technologies in coke industry on aromatic polycyclic hydrocarbons concentrations in ambient air (Pol.). *Med. Pr.*, *29*, 357-364

Buonicore, A.J. (1979) Analyzing organics in air emissions. *Environ. Sci. Technol.*, *13*, 1340-1342

Butlin, H.T. (1892) Three lectures on cancer of the scrotum in chimney-sweep and others. *Br. med. J.*, *ii*, 1341-1346; *iii*, 1-6; *iii*, 66-71

Casto, B.C., Hatch, G.G., Huang, S.L., Huisingh, J.L., Nesnow, S. & Waters, M.D. (1980) *Mutagenic and carcinogenic potency of extracts of diesel and related environmental emissions*: In vitro *mutagenesis and oncogenic transformation*. In: Pepelko, W.E., Danner, R.M. & Clarke, N.A., eds, *Health Effects of Diesel Engine Emissions: Proceedings of an International Symposium*, Cincinnati, OH, US Environmental Protection Agency, pp. 843-860

Casto, B.C., Hatch, G.C., Huang, S.L., Lewtas, J., Nesnow, S. & Waters, M.D. (1981) Mutagenic and carcinogenic potency of extracts of diesel and related environmental emissions: *In vitro* mutagenesis and oncogenic transformation. *Environ. int.*, *5*, 403-409

Claxton, L.D. (1981) Mutagenic and carcinogenic potency of diesel and related environmental emissions: *Salmonella* bioassay. *Environ. int.*, *5*, 389-391

Curren, R.D., Kouri, R.E., Kim, C.M. & Schechtman, L.M. (1980) *Mutagenic and carcinogenic potency of extracts from diesel related environmental emissions: Simultaneous morphological transformation and mutagenesis in BALB/c 3T3 cells*. In: Pepelko, W.E., Danner, R.M. & Clarke, N.A., eds, *Health Effects of Diesel Engine Emissions: Proceedings of an International Symposium*, Cincinnati, OH, US Environmental Protection Agency, pp. 861-897

Curren, R.D., Kouri, R.E., Kim, C.M. & Schechtman, L.M. (1981) Mutagenic and carcinogenic potency of extracts from diesel related environmental emissions: Simultaneous morphological tranformation and mutagenesis in BALB/c 3T3 cells. *Environ. int.*, *5*, 411-415

Davies, G.M. (1977) A mortality study of coke oven workers in two South Wales integrated steelworks. *Br. J. ind. Med.*, *34*, 291-297

Doherty, J.D. & De Carlo, J.A. (1967) Coking practice in the United States compared with some western European practices. *Blast Furnace Steel Plant*, *55*, 141-153

Eisenhut, W., Langer, E. & Meyer, C. (1982) *Determination of PAH pollution at coke works*. In: Cooke, M., Dennis, A. & Fisher, G.L., eds, *Polynuclear Aromatic Hydrocarbons: Physical and Biological Chemistry, 6th International Symposium*, Columbus, OH, Battelle Press, pp. 255-261

Fannick, N., Gonshor, L.T. & Shockley, J., Jr (1972) Exposure to coal tar pitch volatiles at coke ovens. *Am. ind. Hyg. Assoc. J.*, *33*, 461-468

Gillies, A.T. (1983) Experience in controlling airborne pollutant exposure of operatives at a coke oven. *Ann. occup. Hyg.*, *27*, 221-222

Henry, S.A. (1947) Occupational cutaneous cancer attributable to certain chemicals in industry. *Br. med. Bull.*, *4*, 389-401

Henry, S.A., Kennaway, N.M. & Kennaway, E.L. (1931) The incidence of cancer of the bladder and prostate in certain occupations. *J. Hyg.*, *31*, 125-137

Horton, A.W., Tye, R. & Stemmer, K.L. (1963) Experimental carcinogenesis of the lung. Inhalation of gaseous formaldehyde or an aerosol of coal tar by C3H mice. *J. natl Cancer Inst.*, *30*, 31-43

Hueper, W.C. & Payne, W.W. (1960) Carcinogenic studies on petroleum asphalt, cooling oil, and coal tar. *Arch. Pathol.*, *70*, 106-118

Hurley, J.F., Archibald, R. McL., Collings, P.L., Fanning, D.M., Jacobsen, M. & Steele, R.C. (1983) The mortality of coke workers in Britain. *Am. J. ind. Med.*, *4*, 691-704

Jackson, J.O., Warner, P.O. & Mooney, T.F., Jr (1974) Profiles of benzo[*a*]pyrene and coal tar pitch volatiles at and in the immediate vicinity of a coke oven battery. *Am. ind. Hyg. Assoc. J.*, *35*, 276-281

Jungers, R., Burton, R., Claxton, L. & Huisingh, J.L. (1981) *Evaluation of collection and extraction methods for mutagenesis studies on ambient air particulate*. In: Waters, M.D., Sandhu, S.S., Huisingh, J.L., Claxton, L. & Nesnow, S., eds, *Short-Term Bioassays in the Analysis of Complex Environmental Mixtures II*, New York, Plenum Press, pp. 45-65

Kapitulsky, V.B., Kogan, F.M., Filatova, A.S. & Kuzminykh, A.I. (1974) Hygienic characteristics of new large-capacity coke furnace batteries with smokeless charging of stock and dry quenching of coke (Russ.). *Gig. Tr. prof. Zabol.*, *10*, 1-4

Kawahata, K. (1938) Occupational lung cancers occurring in gas generator workers in the steel industry (Ger.). *Gann*, *32*, 369-387

Kawai, M., Matsuyama, T. & Amamoto, H. (1961) A study on occupational lung cancer of the gas producer workers in Yawata iron & steel works (Jpn). *J. Labour Hyg.*, *10*, 5-9

Kawai, M., Amamoto, H. & Harada, K. (1967) Epidemiologic study of occupational lung cancer. *Arch. environ. Health*, *14*, 859-864

Kennaway, E.L. & Kennaway, N.M. (1947) A further study of the incidence of cancer of the lung and larynx. *Br. J. Cancer*, *1*, 260-298

Kennaway, N.M. & Kennaway, E.L. (1936) A study of the incidence of cancer of the lung and larynx. *J. Hyg.*, *36*, 236-267

Kriebel, D., Commoner, B., Bollinger, D., Bronsdon, A., Gold, J. & Henry, J. (1983) Detection of occupational exposure to genotoxic agents with a urinary mutagen assay. *Mutat. Res.*, *108*, 67-79

Kuroda, S. (1937) Occupational pulmonary cancer of generator gas workers. *Ind. Med.*, *6*, 304-306

Lewtas, J., Bradow, R.L., Jungers, R.H., Harris, B.D., Zweidinger, R.B., Cushing, K.M., Gill, B.E. & Albert, R.E. (1981) Mutagenic and carcinogenic potency of extracts of diesel and related environmental emissions: Study design, sample generation, collection, and preparation. *Environ. int.*, *5*, 383-387

Lindstedt, G. & Sollenberg, J. (1982) Polycyclic aromatic hydrocarbons in the occupational environment with special reference to benzo[*a*]pyrene measurements in Swedish industry. *Scand. J. Work Environ. Health*, *8*, 1-19

Lloyd, J.W. (1971) Long-term mortality study of steelworkers. V. Respiratory cancer in coke plant workers. *J. occup. Med.*, *13*, 53-68

Lloyd, J.W. (1980) *Problems of lung cancer mortality in steelworkers*. In: *Luftverunreinigung durch Polycyclische Aromatische Kohlenwasserstoffe, Erfassung und Bewertung* (Air Pollution by Polycyclic Aromatic Hydrocarbons, Overview and Estimation) (*VDI-Berichte no. 358*), Düsseldorf, Verlag des Vereins Deutscher Ingenieure, pp. 237-244

Lloyd, J.W. & Ciocco, A. (1969) Long-term mortality study of steelworkers. I. Methodology. *J. occup. Med.*, *11*, 299-310

Lloyd, J.W., Lundin, F.E., Jr, Redmond, C.K. & Geiser, P.B. (1970) Long-term mortality study of steelworkers. IV. Mortality by work area. *J. occup. Med.*, *12*, 151-157

MacEwen, J.D., Hall III, A. & Scheel, L.D. (1976) *Experimental oncogenesis in rats and mice exposed to coal tar aerosols*. In: *Proceedings of the Seventh Annual Conference on Environmental Toxicology, 13-15 October 1976, Dayton, OH* (*AMRL Technical Report No. 76-125*), Wright-Patterson Air Force Base, OH, Aerospace Medical Research Laboratory

Mašek, V. (1971) Benzo[*a*]pyrene in the workplace atmosphere of coal and pitch coking plants. *J. occup. Med.*, *13*, 193-198

Mašek, V. (1972a) Aldehydes in the air at workplaces in coal and pitch coking plants. *Staub.-Reinhalt Luft*, *32*, 26-28

Mašek, V. (1972b) New results on the properties of the dusts of a coke plant. II. Occurrence of arsenic in the air dust of carbonisation (Ger.). *Zentralbl. Arbeitsmed. Arbeitsschutz*, *22*, 69-74

Mašek, V. (1974) Hydrogen cyanide in the air of coking plant working environment (Ger.). *Zentralbl. Arbeitsmed. Arbeitsschutz*, *24*, 101-105

Mašek, V. (1978) Benzo[*a*]pyrene determination in airborne soot from metallurgical factories (Ger.). *Zentralbl. Arbeitsmed. Arbeitsschutz*, *28*, 168-170

Mašek, V. (1979) Emissions generated during coke pressing from the chambers and during moist quenching (Ger.). *Zentralbl. Arbeitsmed. Arbeitsschutz*, *29*, 103-109

Mašek, V. (1980) Quality of solid particles emitted into the atmosphere in wet quenching of coke (Czech.). *Cesk. Hyg.*, *25*, 144-150

Mašek, V. & Neiser, J. (1973) Gas chromatographic analysis of specific pyridines in the air of coking plants (Ger.). *Zentralbl. Arbeitsmed. Arbeitsschutz*, *23*, 337-339

Mazumdar, S., Redmond, C., Sollecito, W. & Sussman, N. (1975) An epidemiological study of exposure to coal tar pitch volatiles among coke oven workers. *J. Air Pollut. Control Assoc.*, *25*, 382-389

McGannon, H.E., ed. (1971) *The Making, Shaping and Treating of Steel*, Pittsburgh, PA, United States Steel Corp., pp. 1-10, 105-106

Menck, H.R. & Henderson, B.E. (1976) Occupational differences in rates of lung cancer. *J. occup. Med.*, *18*, 797-801

Mitchell, A.D., Evans, E.L., Jotz, M.M., Riccio, E.S., Mortelmans, K.E. & Simmon, V.F. (1980) *Mutagenic and carcinogenic potency of extracts of diesel and related environmental emissions:* In vitro *mutagenesis and DNA damage*. In: Pepelko, W.E., Danner, R.M. & Clarke, N.A., eds, *Health Effects of Diesel Engine Emissions: Proceedings of an International Symposium*, Cincinnati, OH, US Environmental Protection Agency, pp. 810-842

Mitchell, A.D., Evans, E.L., Jotz, M.M., Riccio, E.S., Mortelmans, K.E. & Simmon, V.F. (1981) Mutagenic and carcinogenic potency of extracts of diesel and related environmental emissions: *In vitro* mutagenesis and DNA damage. *Environ. int.*, *5*, 393-401

Møller, M. & Dybing, E. (1980) Mutagenicity studies with urine concentrates from coke plant workers. *Scand. J. Work Environ. Health*, *6*, 216-220

National Institute for Occupational Safety and Health (1973) *Criteria for a Recommended Standard... Occupational Exposure to Coke Oven Emissions* (*HSM 73-11016*), Washington DC, US Government Printing Office

National Institute for Occupational Safety and Health (1975) *An Industrial Hygiene Survey of the Bethlehem Steel Corp. (Burns Harbor Facility) Coke Side Emission Collecting Shed* (*HHE 75-32*), Cincinnati, OH

National Institute for Occupational Safety and Health (1978) *Recommended Health and Safety Guidelines for Coal Gasification Pilot Plants* (*DHEW (NIOSH) Publ. No. 78-120*), Cincinnati, OH

Nesnow, S., Triplett, L.L. & Slaga, T.J. (1982) Comparative tumor-initiating activity of complex mixtures from environmental particulate emissions on SENCAR Mouse skin. *J. natl Cancer Inst.*, *68*, 829-834

Nesnow, S., Triplett, L.L. & Slaga, T.J. (1983) Mouse skin tumor initiation - promotion and complete carcinogenesis bioassays: Mechanisms and biological activities of emission samples. *Environ. Health Perspect.*, *47*, 255-268

Nilsen, A.M., Madslien, O., Mylius, E.A. & Gullvåg, B.M. (1984) Alveolar macrophages (AM) from expectorates samples. A stress signal from occupational pollution. *Bull. environ. Contam. Toxicol.* (in press)

O'Donovan, W.J. (1920) Epitheliomatous ulceration among tar workers. *Br. J. Dermatol. Syph.*, *32*, 215-229

O'Donovan, W.J. (1928) *Cancer of the skin due to occupation*. In: *International Conference on Cancer*, London, Fowler Wright, pp. 292-303

Radford, E.P. (1976) Cancer mortality in the steel industry. *Ann. N.Y. Acad. Sci.*, *271*, 228-238

Redmond, C.K., Ciocco, A., Lloyd, J.W. & Rush, H.W. (1972) Long-term mortality study of steelworkers. VI. Mortality from malignant neoplasms among coke oven workers. *J. occup. Med.*, *14*, 621-629

Redmond, C.K., Strobino, B.R. & Cypess, R.H. (1976) Cancer experience among coke by-product workers. *Ann. N.Y. Acad. Sci.*, *217*, 102-115

Redmond, C.K., Wieand, H.S., Rockette, H.E., Sass, R. & Weinberg, G. (1981) *Long-term Mortality Experience of Steelworkers* (*DHHS (NIOSH) Publ. No. 81-120*), Cincinnati, OH, National Institute for Occupational Safety & Health

Reid, D.D. & Buck, C. (1956) Cancer in coking plant workers. *Br. J. ind. Med.*, *13*, 265-269

Ross, P. (1948) Occupational skin lesions due to pitch and tar. *Br. med. J.*, *ii*, 369-374

Sakabe, H., Tsuchiya, K., Takekura, N., Nomura, S., Koshi, S., Takemoto, K., Matsushita, H. & Matsuo, Y. (1975) Lung cancer among coke oven workers. A report to Labour Standard Bureau, Ministry of Labour, Japan. *Ind. Health*, *13*, 57-68

Selikoff, I.J. & Hammond, E.C. (1971) Asbestos exposure to coke oven operators. *J. occup. Med.*, *13*, 496-497

Szyja, J. (1977) Studies on benzo[*a*]pyrene content in the air at coke factories and in the blood and urine of men exposed to coke ovens (Ger.). *Z. gesamte Hyg.*, *23*, 440-442

Tanimura, H. (1968) Benzo[a]pyrene in an iron and steel works. *Arch. environ. Health*, *17*, 172-177

Tye, R. & Stemmer, K.L. (1967) Experimental carcinogenesis of the lung. II. Influence of phenols in the production of carcinoma. *J. natl Cancer Inst.*, *39*, 175-186

US Occupational Safety & Health Administration (1976) Occupational safety and health standards. Exposure to coke oven emissions. *Fed. Regist.*, *41*, 46742-46790

Walker, D.D., Archibald, R.M. & Attfield, M.D. (1971) Bronchitis in men employed in the coke industry. *Br. J. ind. Med.*, *28*, 358-363

Yanysheva, N.Y., Kireeva, J.S. & Serzhantova, N.N. (1962) 3,4-Benzo[*a*]pyrene in discharges of coke chemical plants and in atmospheric air pollution (Russ.). *Gig. Sanit.*, *27*, 3-7

IRON AND STEEL FOUNDING

The iron and steel founding industry, as referred to in this monograph, begins with the melting of alloys for the purposes of casting and ends with the fettling of castings.

1. Historical Perspectives

This subject was reviewed by Beeley (1972).

Metal founding is one of the oldest of all industries, both ancient and mediaeval history offering examples of the manufacture and use of castings. From simple axeheads poured from copper in open moulds some 5000 years ago, founding in the pre-Christian world developed to a point at which elaborate bronze statuary could be produced in two-piece and cored moulds. By the end of the mediaeval period, decorated bronze and pewter castings had begun to be used in European churches and for domestic purposes, while cast iron made its first appearance in the shapes of cannon shot and grave slabs. Moulding boxes and sand were in use by the sixteenth century.

In the seventeenth century, interest arose in the structure of cast iron and in factors influencing the production of white, grey and mottled irons. The widespread adoption of cast iron as an engineering material awaited the success of smelting iron in coke blast furnaces in 1709 by Abraham Darby. This enabled the massive use of cast iron in construction. Bulk supplies of steel became available for the foundry industry after development of the Bessemer converter in 1856.

Historically, metal casting was looked upon as an art, a view which persisted until well into the present century. The foundry industry grew around the skills of the moulder and the patternmaker who were able to produce complex forms in various sizes. However, it was the mechanization of production methods that permitted the development of the modern foundry industry. Earlier it was common practice for the moulder to receive the pattern, determine the casting method and carry out his own sand preparation, moulding and pouring; he might even melt the alloy and fettle his own castings. During the last decades, sand mixers, belt conveyors, automatic moulding lines and mechanized fettling processes have replaced most of the older operations.

While silica sands and clays have been the principal moulding constituents in which metals have been cast from early history, little was written on the subject until the beginning of the twentieth century. At that time, two mixtures were in use: the moulding sand, bonded solely by natural clay, and the core sand, which was coarser in texture and contained binders in addition to clay. Molasses, starch, flour and linseed oil were dry binders used in the early bonding of sand (Woodliff, 1971). The addition of coal powder to foundry sand has been practised for over a century (Radia, 1977). The carbonaceous additives, whose volatile matter was distilled off by the heat of the casting metal, included pitch, sea-coal, rosin, gilsonite and various fuel oils (Woodliff, 1971).

With the advent of synthetic binder materials, cores were decreasingly bonded with linseed oil; drying oils derived from coal or petroleum distillation were substituted. Iron oxide used directly in the sand mixture prevented veining and penetration of the cast metal. In about 1940, wood flour was introduced to moulding and core sands in order to control collapsibility. Other cellulosic materials - ground straw, oat hulls and corn flour - were soon available for thermal expansion cushioning in moulding sands. Mould paints and blackings such as graphite, coal or zircon were used for treating sand surfaces before casting (Woodliff, 1971).

Foundry sands have traditionally contained organic binders, as mentioned above. The shortage of resins and oils created by the Second World War stimulated investigation of other materials. In 1948, a British patent covering a process for producing cores and moulds with sodium silicate was published. Solidification of the sand mixture was accomplished by gassing with carbon dioxide. The carbon dioxide process was the first in which cores were produced without heat (Woodliff, 1971).

The shell-moulding process was developed in Germany by Johannes Croning in the 1940s. Large-scale use of shell moulding began in about 1950, and since that time there has been a rapid increase in its use. The early processes utilized modified phenol formaldehyde resins. In the late 1950s, urea resins combined with furfuryl alcohol were used in the first self-setting, no-bake core process (Woodliff, 1971).

In 1965, an alkyd-isocyanate, no-bake process was introduced. Usage of these isocyanate binders has grown steadily so that from their introduction up to 1980 over 200 000 tonnes of urethane binders have been consumed by the US foundry industry, with an equal amount being used elsewhere in the world (Toeniskoetter, 1981). Overall, the use of chemically bonded sand has increased from less than 100 000 tonnes in 1958 to over 3 million tonnes in 1978 (Goss, 1980).

Substitutes for silica sand were first introduced in a Norwegian steel foundry in 1927, where crushed olivine was used as a moulding sand. In the USA, commercial production of steel castings in olivine moulds began in 1953. Thereafter, a large number of foundries in many countries were using olivine, zircon or chromite sand for limited or special applications (Tubich, 1964). For example, in 1975, ferrous foundries in the USA used 10 million tonnes of silica sand, 80 000 tonnes of chromite sand, 50 000 tonnes of zircon sand and 45 000 tonnes of olivine sand (Rassenfoss, 1977).

Other recent changes include increased mechanization and automation of foundries and the introduction of a wide range of new chemical binders and new alloys used for casting. Cast iron may contain some manganese and 2.5-4.5% carbon, whereas steel also sometimes contains significant amounts of chromium and nickel but much less carbon (0.5%) (National Institute for Occupational Safety & Health, 1978).

In 1981, world production of iron and steel castings exceeded 77 million tonnes (Table 1). There are about 10 000 foundries worldwide, employing approximately 2 million workers, as estimated from various sources and on the basis of production figures (Table 2). The USA and the USSR together account for about one half of the production of ferrous castings. Iron and steel castings are mainly used in construction and engineering applications, such as machine and motor parts, process equipment, pipes, pumps and valves.

Many foundries are incorporated in other metal industries that include castings in their own finished products. Although most foundries base their activities on a limited range of alloys, e.g., cast iron or steel, copper or aluminium alloys, others make use of several of these

Table 1. Summary of world production of castings in 1981 (1000 tonnes)[a]

Country	Iron castings	Nodular iron castings	Steel castings	Malleable iron castings	Total
Argentina (1980)	172.7	25.2	18.4	8.4	224.7
Australia[b]	415.0	63.0	72.0	16.0	566.0
Austria	119.4	39.0	31.9	13.5	203.8
Belgium	127.8	7.3	72.9	-	208.0
Brazil	900.7	272.8	140.5	46.9	1 360.9
Canada	521.7	210.2	154.1	15.1	901.1
Chile (1977)	30.0	0.8	13.0	-	43.8
Czechoslovakia	1 069.1	23.9	363.6	32.1	1 488.7
Denmark[b]	65.0	6.0	6.0	5.0	82.0
Egypt	59.4	-	5.9	-	65.3
Federal Republic of Germany	2 509.1	746.5	280.7	147.2	3 683.5
Finland	79.9	19.0	19.3	2.0	120.2
France	1 257.1	770.4	201.6	58.6	2 287.7
German Democratic Republic (1980)	976.6	80.8	233.2	37.2	1 327.8
Hungary	256.9	1.0	54.7	7.3	319.9
India	240.0	7.4	74.0	19.0	340.4
Israel	21.0	1.5	7.5	3.0	33.0
Italy	1 404.8	177.4	115.3	50.1	1 747.6
Japan	3 949.0	1 548.5	682.7	299.3	6 479.5
Luxembourg (1980)	39.0	23.8	-	-	62.3
Mexico (1980)	737.9	43.1	77.2	11.0	869.2
The Netherlands	233.7	24.1	4.3	8.8	270.9
New Zealand (1980)	21.6	-	-	-	21.6
Norway	71.5	16.5	5.6	12.3	105.9
People's Republic of China	3 971.0	242.0	683.0	263.0	5 159.0
Peru (1979)	25.0	5.0	40.0	3.0	73.0
Philippines	65.0[c]	-	37.0	-	102.0
Poland (1978)	1 946.0[c]	-	355.0	-	2 301.0
Portugal[b]	46.0	12.0	12.0	14.0	84.0
Republic of Korea	410.0	94.0	88.0	28.0	620.0
Roumania	1 182.5	32.1	357.2	11.4	1 583.2
Singapore (1979)	24.0[c]	-	6.0	-	30.0
South Africa	297.7	27.0	141.7	33.9	500.3
Spain (1980)	550.0	140.0	95.0	36.0	821.0
Sweden	223.0	37.0	13.0	7.0	280.0
Switzerland	237.4[d]	-	8.6	0.1	246.1
Taiwan	354.5	27.3	30.4	33.1	445.3
Turkey	225.0	8.2	45.0	6.5	284.7
United Kingdom	1 268.8	244.3	150.0	131.7	1 794.8
USA	8 826.3	1 997.0	1 589.7	381.9	12 794.9
USSR (1980)	17 959.0	328.0	7 745.0	899.0	26 931.0
Yugoslavia(1979)	472.5	34.2	70.3	30.2	607.2
Zambia	1.0	-	27.6	-	28.6
Total	53 363.6	7 335.8	14 128.9	2 671.6	77 499.9

[a]From Comité des Associations Européennes de Fonderie
[b]Estimated
[c]Includes nodular and malleable iron castings
[d]All nodular iron castings

Table 2. Employment in iron and steel foundries[a]

Country	No. of foundries	Approximate no. of workers	Year
Austria	55	8 902[b]	1982
Belgium	98	6 407	1981
Brazil[c]	268	62 223	1974
Denmark	23	2 757	1979
Federal Republic of Germany	537	83 482	1982
Finland	38	3 762	1982
France	344	57 490	1982
Italy	537	43 920	1982
Japan[d]	1 225	76 610	1980
The Netherlands	50	3 858	1982
Norway	23	2 516	1982
Portugal	123[e]	7 131[e]	1982
Sweden	80	6 700	1982
Switzerland	38	6 218	1982
United Kingdom	563	56 976	1982
USA[f]	1 971	307 200	1975
USSR[g]	-	484 000	1973

[a]From Comité des Associations Européennes de Fonderie (1983), unless otherwise indicated
[b]Includes non-ferrous metal castings
[c]Estimated
[d]Schneider (1976)
[e]Standke & Buchen (1982)
[f]Shestopal (1976)
[g]National Institute for Occupational Safety & Health (1978)

materials side by side. The temporal development of the foundry industry has been uneven, and in most countries all technological stages from manual floor moulding to fully automated production are still in operation.

During recent years, there has been a trend toward fewer but larger foundries with high productivity in tonnes/employee; but most foundries still employ fewer than 50 workers. In 1975, about 45% of foundry employees in the USA produced iron castings, 16% produced steel castings and 39% produced non-ferrous castings (National Institute for Occupational Safety & Health, 1978). About one half of the employees worked in foundries employing more than 250 people. These large foundries produce most of the castings, and in them workers have a single job assignment; approximately one-third of the workers are moulders and coremakers, one-third are fettlers, and the remainder furnacemen, crane drivers, supervisors, patternmakers, maintenance workers and others. In small shops, a single worker may be responsible for several tasks.

2. Description of the Industry

This subject was reviewed by Beeley (1972).

2.1 Introduction

Foundries produce shaped castings from remelted metal ingots and scrap. Shaped castings are distinguished from ingots and other cast forms which are produced in iron and steel works outside the foundry industry. Although founding work is assumed to start from the remelting of ingots and scrap and to end with the fettling of castings, the industry is often so integrated that the division is less obvious. Machine shops are not normally a part of castings production areas; however, simple and accessory machining may be carried out, and they may be part of foundry operations in small operations.

Iron and steel foundries generally comprise the following basic sections (see Fig. 1):

- patternmaking
- moulding
- coremaking
- melting and pouring
- shake-out
- fettling.

Fig. 1. Iron foundry processes, indicating emission sources[a]

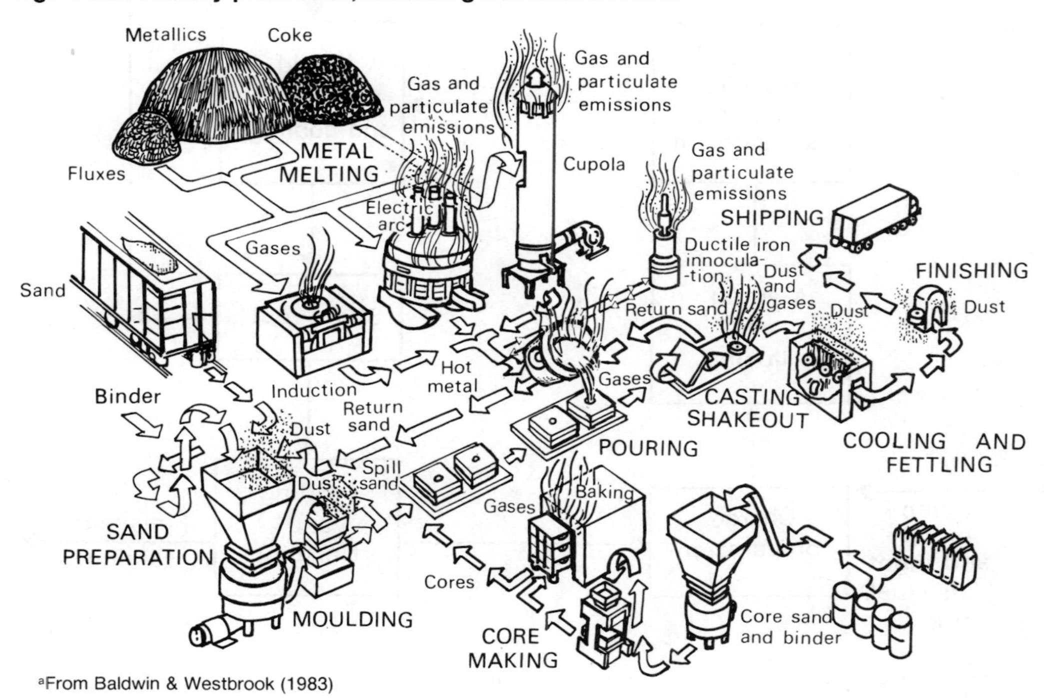

[a]From Baldwin & Westbrook (1983)

Figure 2 illustrates the flow of operations in a sand cast foundry. In old or small foundries different processes may be carried out simultaneously or consecutively in the same workshop.

Fig. 2. Typical flow sheet of operations in a sand-cast foundry[a]

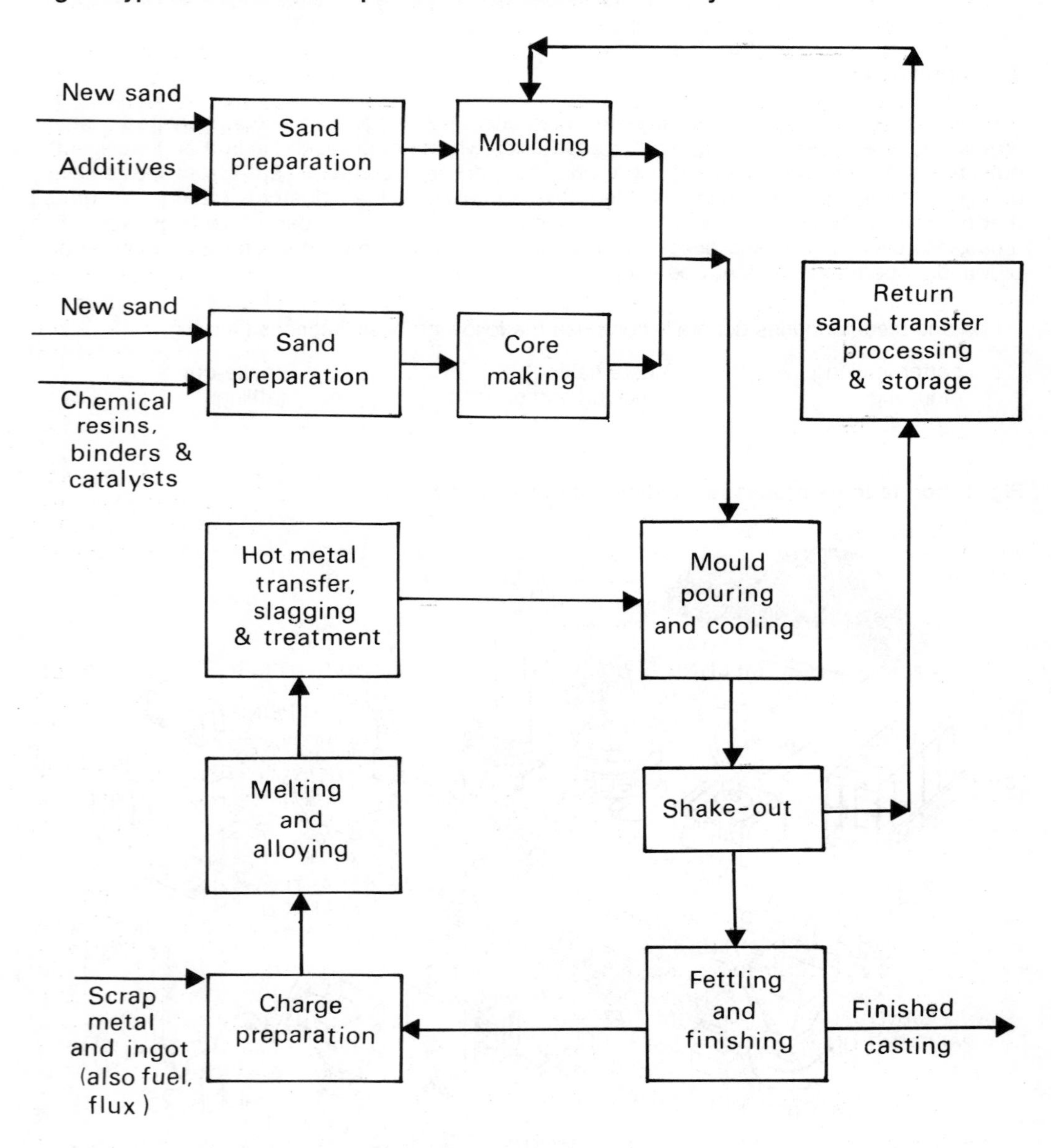

[a]From National Institute for Occupational Safety & Health (1978)

2.2 Foundry operations

(*a*) *Patternmaking*

Pattern materials and designs depend upon the method of casting production and the shapes desired. Patterns can range from simple wooden objects, intended for a single use, to elaborate metal assemblies, complete with gating and feeding systems and suitable for mass production by automatic moulding lines. A pattern shop is usually located in a separate department at some distance from the foundry because it is necessary to keep the machinery free from sand dust. Wood- and metal-working machines, saws, planes and lathes are to be found in the pattern shop, in addition to carpenters' benches at which hand tools are used. Instead of wood, patterns can be manufactured from reinforced epoxy or polyurethane plastics. Glues, paints and various solvents are frequently used. Occupational exposures of patternmakers clearly differ from those of other workers in iron and steel foundries.

Expendable patterns are made from fusible or destructible material as a part of the production sequence for each casting. In the lost wax process, patterns are produced by gravity pouring or by pressure injection. Natural or synthetic waxes melt within the range of 55-90°C in the moulds, which are then ready for high-precision or investment casting. Polystyrene foam patterns, which vapourize during casting, are sometimes employed for mass produced components (Heine, 1977).

(*b*) *Moulding and coremaking*

Foundry sand passes through several production stages, namely preparation and distribution, mould and core production, casting, shake-out and reclamation. The requirements of the materials are determined by the moulding and casting conditions. Strength, flowability, refractoriness, durability and collapsibility ('knock-out' properties) are the most important qualities.

(i) *Sand and binder materials*

Moulding sands fall into two broad categories according to the type of base sand employed. Naturally bonded sands are those in which the refractory grains are associated in their deposits with the clay needed for moulding. Such sands often develop sufficient properties with the addition of water alone, but their high clay content may reduce refractoriness and gas permeability. Synthetic sands are based mainly on silica sands or crushed stone containing no binder in the natural state, the strength properties being achieved by separate additions. The critical characteristics of the base sand are its chemical composition, mechanical grading and grain shape.

Most moulding sands are based upon the mineral quartz, which is both geologically abundant and refractory to about 1700°C. Phase transformations to cristobalite or tridymite may occur at lower temperatures (Gerhardsson, 1976). For maximum refractoriness, the quartz sand should not contain fusible minerals, notably the feldspars, mica or alkali fluxes. In the production of heavy steel castings, silica sand may be replaced by still more stable materials such as zircon, olivine or high-alumina chamotte (a synthetic alumino-silicate). Grain sizes of foundry sands fall mainly within the range of 0.1-1.0 mm. Grain shape and surface texture are determined by the geological origin, rounded grains being associated with natural sands, whereas angular shapes are formed from crushed stone (Beeley, 1972).

The function of the binder is to produce cohesion between the refractory grains in the green or dried state. Many substances possess bonding qualities: clays, organic oils and resins, cereals and water-glass silicates may be used singly or in combination. Clay-bonded sands are distinguished by the fact that they can be recirculated in closed systems and the bond regenerated by necessary additions. The action of most other binders is irreversible, and the moulding material is discarded after a single production cycle. Adequate strength can be derived with contents as low as 3-5%. Most ferrous castings are produced in green-sand moulds. The term green sand implies that the binders include clay.

Green-sand moulds also contain organic additives such as cereal, dextrin, starch, wood flour, pitch or pulverized coal dust. These agents are present in amounts of up to 6% in the sand and provide a reducing atmosphere inside the mould during casting. As the metal is poured, these materials partially decompose, and hydrogen, carbon monoxide, carbon dioxide, water vapour and volatile hydrocarbons are driven off into the foundry atmosphere. A graphitic layer of lustrous carbon is formed on the surface of sand grains, preventing sand-metal reactions. Coal-dust replacements are sold as pure or clay-blended mixtures as well as in liquid form. These products are based mainly on the following substances, used singly or in combination (Radia, 1977):

Synthetic polymers: polystyrene, polyethylene, polypropylene and polymeric petroleum products in solid form or in aqueous dispersions

Oils: products of coal or petroleum distillation, which comprise complex mixtures of heavy hydrocarbons, aromatic compounds and naphthenes

Asphalts: natural asphalts such as gilsonite or synthetic asphalts (bitumens), which are produced in petroleum refining

Pitches and bitumens: coal-tar pitch and petroleum bitumen in pulverized or pelleted form

Some of the principal core binders are listed in Table 3 (Emory *et al.*, 1978). These are broadly categorized with regard to the method of curing - by oven baking, by heating inside the core box, by gassing or by addition of a polymerization catalyst.

Table 3. Some types of organic sand binders[a]

Oils
Core oils
Oil-oxygen no-bake
Hot box
Urea-formaldehyde
Phenol-formaldehyde
Furan
Modified furan
Urea-formaldehyde/furfuryl alcohol
Phenol-formaldehyde/furfuryl alcohol
Phenol-formaldehyde/urea-formaldehyde
Urethane no-bake
Alkyd isocyanate
Phenolic isocyanate
Polyester urethane
Acid no-bake
Furan with:
Phosphoric acid
Toluenesulphonic acid
Benzenesulphonic acid
Phenol-formaldehyde

[a]Adapted from Emory *et al.* (1978)

The oleoresinous binders are combinations of vegetable oils, petroleum polymers and organic solvents. The resins polymerize in the presence of oxygen to produce a compound of high molecular weight which acts as a sand binder.

Urea-formaldehyde (UF) resins represent a second type of baked core binder. They are usually blended with oleoresinous or phenol-formaldehyde (PF) resins to obtain the desired hot strength, knock-out properties and curing times. Phenol-formaldehyde resins are currently in use in the oven baking processes as well as in shell and hot-box processes. Hot-box binders are those resins that polymerize rapidly in the presence of acidic chemicals and heat. The original hot-box binders were developed by modifying urea-formaldehyde resins with the addition of 20-45% furfuryl alcohol (FA). This mixture was commonly referred to as a furan resin. The furan resins were then modified with phenol to produce UF/FA, PF/FA and PF/UF resins. The shell-moulding process uses premixed sands which contain 3-6% of phenolic resin, hexamethylenetetramine as a catalyst, and silicones or stearates as parting compounds (Tubich *et al.*, 1960).

The polyurethane cold-box processes employ two component resins. In the presence of an amine catalyst, the hydroxyl groups of the liquid phenolic resin combine with the isocyanate groups of the pre-polymerized 4,4′-methylenediphenyl diisocyanate (MDI) resin to form a solid polyurethane. Cold-setting binders are also commonly based on phenolic or furan resins in which the setting reaction is catalysed by the addition of phosphoric, toluenesulphonic or benzenesulphonic acid at the time of mixing. The rate of hardening is high and can be controlled by the amount of the acid accelerator. A more recent development of the cold-setting furan method involves the use of sulphur dioxide and methylethylketone peroxide as hardening agents (Gardziella, 1982).

Organic binders decompose with the heat during casting. The total emission of gases and vapours is lower in comparison to those from conventional green sand, but their composition is highly dependent on the type of resin system. Heating beyond the curing temperature produces charring followed by complete oxidation and burnout. Urea-formaldehyde resins possess the lowest breakdown temperatures, usually in the range 180-230°C. Linseed oil breaks down rapidly at 450-500°C and phenol-formaldehyde, hot-box and cold-box resins at a slightly lower temperature (Beeley, 1972).

An inorganic binder of importance is sodium silicate, which is hardened by gassing with carbon dioxide. Passage of carbon dioxide through the sodium silicate-impregnated sand produces colloidal silica gel, which hardens and forms a bond between the sand grains. The method can be applied to both moulding and core sands. Sodium silicate and ethyl silicate are used with other chemical hardening agents as a cold-setting binder (Beeley, 1972).

The use of cement as a binder forms the basis of a distinctive form of moulding practice. Silica sand containing 11% Portland cement and 6% water exhibits low strength initially, but develops high dry strength during drying and setting. Either hand or machine moulding can be employed and the process has found application in both ferrous and non-ferrous founding (Beeley, 1972).

The need to provide reasonable gas permeability combined with low porosity and a smooth mould surface is resolved in many cases by the use of refractory mould dressings or blackings. A coating may be applied either as a liquid wash by brush or spray gun, or as dry solids by dusting. Typical refractories include graphite, coke, silica, alumina, magnesia, chromite, zircon, talc, mica, sillimanite, kyanite and chamotte (Beeley, 1972; Lukacek & Heine, 1982). Carbonaceous dressings are widely favoured in iron founding, but other parting materials are also essential for permanent mould and die casting. The liquid carrier may contain water, alcohols or chlorinated hydrocarbons. As a fast evaporating medium, 1,1,1-trichloroethane is sometimes used. However, isopropanol is most commonly used because it can be burned off easily.

(ii) *Moulding operations*

The plant for sand preparation is normally provided with facilities for storing, mixing and transporting new and reclaimed sand, bonding materials and special additives. Batch mixers with rotating rollers, ploughs or paddles are used in smaller and older foundries. The most primitive system is based upon hand mixing with a shovel. At intervals the sand heap is reconditioned with additions of new sand and water. In contrast to such simple methods are the fully integrated sand-processing plants with sand reclamation, screening and aerating. Processes of pneumatic or wet scrubbing and separation are employed in removing clay, but thermal reclamation, involving high-temperature oxidation, is needed for complete elimination of carbonaceous residues in resin sands. In modern plants, closely controlled mixtures are produced from new and reclaimed sand, binders and additives. Bulk materials are fed directly into the mixers and mills in measured quantities without intermediate handling, and, thereafter, delivered by skips or belt conveyors to the moulding stations. Moulding machines and sandslingers are served by feed hoppers. After casting, sand from shake-out is returned to the sand plant. Sand systems of this type can be highly automated, with centralized control of the entire process.

Two main operations are carried out in the foundry *per se*: sand moulding and casting of the metal. In many foundries, the cores and patterns are made in separate workrooms. The moulders make the moulds on a bench if the castings are small and on the foundry floor or in a pit when the castings are medium or large. The object of moulding is to produce accurate parts able to withstand lifting and handling and to contain the ferrostatic pressures in casting. Mould strength depends partly on effective sand compaction, reinforcements and mechanical support derived from the moulding box or flask. The general aim in compaction is to bring the bulk density of the sand from its loose value of about 0.6-0.8 g/cm^3 to a density of 1.6-1.8 g/cm^3 when fully packed (Beeley, 1972). Hand ramming is usually adopted for jobbing production where small quantity requirements or unsuitable size preclude the use of moulding machines. Vibration produced by pneumatic tools aids the manual ramming.

Machine moulding enables moulds to be made in large quantities. In most machines, the table carrying the pattern assembly is repeatedly jolted against a stop at the time that the moulding box is being filled with sand from a sandslinger. Jolting produces a mould density which can be increased further by squeezing. Although moulding machines are commonly operated under manual control, automation is provided on modern equipment with preset timing of production sequences. Such machines are followed in varying degrees by automatic coring, closing and handling through the casting and shake-out stages. The systems are based on either the rotatory or in-line principle.

Some moulds are dried before closing and casting. Stoving is most commonly carried out at 200-400°C, above which loss of combined water is detrimental to clay binders. Drying equipment includes fuel-fired stoves and gas torches. Electric or microwave mould driers are also available.

(iii) *Coremaking operations*

Cores are used as inserts in moulds to form casting features such as cavities and holes. Sand cores are compacted by hand ramming, jolting and squeezing operations analogous to those in moulding. The most distinctive technique is, however, the use of coreblowers, which blow the sand into the core box at high velocity and produce small and medium size cores at high rate.

Thermal curing of linseed oil or synthetic baked resins requires heating. Core-stoves are operated at 180-250°C; in addition, air circulation is needed to supply oxygen to the core surfaces and to eliminate moisture and fumes generated in baking (Beeley, 1972). Some of the resin binders are hardened by chemical additives in shell-moulding, hot-box, cold-box and carbon dioxide processes. Although they were originally conceived as coremaking processes, most of the techniques are also used in mould production. The hot-box and shell processes use heated core boxes made of metal. The core is blown in a machine, and curing proceeds at a rate sufficient for ejection within a few seconds. The curing is completed by residual heat in the core sand.

(*c*) *Melting and pouring*

Most of the cast iron produced in foundries is melted in cupolas. When the cupola is charged at the top of the shaft with pig iron, scrap, coke, limestone and fluxes, combustion of the coke melts the alloy. The cupola can be further equipped to provide hot blasting, oxygen enrichment and water cooling. During the melting process, combustion gases rise through the full stack, the molten metal descends through the coke bed and is tapped out at the bottom of the furnace.

Electric-induction furnaces are increasingly used for melting iron and steel. They allow flexible control of alloy temperature and composition. In the absence of fuel combustion, electric furnaces do not evolve much gaseous emissions, except when oxygen is occasionally lanced into the melt for carbon removal. High-frequency induction furnaces are refractory-lined structures with a capacity of several tonnes. Vacuum melting, which may be applied to induction furnaces extracts dissolved gases from the alloy and thus prevents the spread of fumes into the work environment.

Electric arc furnaces are predominant in steel founding. Their capacities range from a few tonnes to about 200 tonnes. Melting of the charge is produced by an electric arc of over 3000°C generated by carbon electrodes through which current is passed. Fuel-fired hearth furnaces, including rotary and tilting reverberatory furnaces, are sometimes used for cast-iron production. The cupola-converter combination and the open-hearth furnace account for a proportion of steel-castings production, especially when a foundry is connected with an adjacent steel works. In 1975, there were 1405 cupolas, 4688 electric furnaces, 49 open hearths and 1746 reverberatory furnaces in the USA and Canada. About 80% of cast iron was produced in cupolas (Miske, 1978).

In general, foundry melting is carried out on a batch basis, the exception being the continuous cupola process. In some cases, the duplex practice is used, which involves separate furnaces for successive stages. A holding furnace provides a reservoir of molten metal at the pouring temperature.

The furnace charge may consist of ingots, alloys and scrap from the fettling and machine shops or outside sources. Melting materials are customarily stored in piles and heaps in the foundry yard. To reduce scrap to sizes suitable for the furnace, flame cutting or mechanical breaking may be applied before charging. Alloy additions are purchased as lumps of pure metal or ferroalloys such as ferromanganese, ferrochromium, ferrosilicon and ferrophosphorus. Calcium compounds - calcium silicide, calcium carbide, calcium fluoride and calcium carbonate

- are also used for melt treatment. Magnesium metal is added to cast iron in the production of nodular iron. The scrap may contain lead, zinc, cadmium and other undesirable metal constituents. During melting they are evaporated, together with pyrolysis products of any oil, grease, plastics or rubber present in the charge (Bates & Scheel, 1974).

The final stage of melting often includes purification, deoxidation, degassing, inoculation or grain refinement of the alloy. In such treatments the reagents are added into the furnace or ladle by plunging, injection or similar techniques. The appropriate tapping temperature depends on the expected rate of cooling in the ladle relative to the timing of casting operations. Cast irons are poured at about 1400°C and high-alloy steels at somewhat higher temperatures.

The molten metal is transported to the casting area in ladles which are of the lip type or are fitted with a nozzle and stopper for bottom pouring. During production of heavy castings, the ladles are supported by cranes, while smaller quantities can be dispensed from underslung hoists or hand shanks. The organization of pouring is regarded as the crucial step in founding, since little time is available for casting operations, and an error in pouring rate or temperature may deteriorate the mould and casting. The layout of the casting floor is based on a fixed point to which moulds are fed by a roller conveyer, or moulds are laid out on the floor. A necessary adjunct is a ladle preheating system that ensures that the repaired refractory lining is thoroughly dried out and minimizes the temperature drop of the metal from furnace to ladle. Various ancillary tasks during and after pouring include addition of insulating material, feeding of exothermic materials, removal of relief cores and lighting of gas evolving from the hot mould. The shake-out section is provided with various mechanical aids: pneumatic tools, hammers, punchout rams, vibratory tables and grids. Equipment for sand collection and dust exhaust is also located in this section of the foundry.

(*d*) *Fettling*

After shake-out, the remaining work in producing the finished casting is carried out in the fettling shop. Basic operations are initial cleaning for removing adherent sand residues, separation of feeder heads and runners (excess metal), and final dressing. Other operations often integrated with these processes are inspection, rectification, heat treatment and primer painting.

The rough cleaning is first done manually or mechanically with crowbars or pneumatic picks to loosen residual sand. In abrasive blasting, individual castings or batch loads of smaller pieces are cleaned by jets of abrasive particles. The abrasives are energized either by compressed air or by centrifugal impeller. Quartz sand, steel shot, cut wire or slag pellets are commonly used. Heavy castings are normally processed in a fixed chamber, the shot nozzle being manipulated directly by the operator, clad in a protective suit, or remotely controlled from outside. Smaller castings are abrasive-blasted in a glove box. In hydraulic cleaning systems, the castings are exposed to high-pressure water jets, a method which eliminates dust. Other methods of cleaning include barrel tumbling and wire brushing.

Many alternative operations are used for the removal of feeders, risers, gates and sprues from the casting. These extend from simple knocking off or prizing away, through sawing, grinding or other machining methods to flame cutting. The oxyacetylene cutting torch is capable of sufficient penetration through carbon and low-alloy steels, but iron powder dispensed into the flame may be needed for stainless-steels and for cutting through large sections. Another process involves use of an electric arc combined with a compressed-air jet to cut the metal. Grinding operations are carried out on swing-frame, pedestal or portable machines. The first of these is suitable for heavy-duty grinding on large castings. Pedestal or bench grinders are used when castings can be lifted manually. The most universal tool for general dressing is the portable grinder. The disc grinders are powered either electrically or by compressed air and may be axially mounted wheels or angled shafts. The abrasives are often made of silicon carbide and alumina bonded with ceramics or synthetic resins.

Castings cannot be produced regularly without defects. Rectification or minor repairs may be needed for the restoration of standard properties or for the improvement of appearance. Fine cracks are detected by means of colour penetrant, X-ray or ultrasonics. Gas and arc welding are most commonly used for repairs. The defects are first removed by grinding, chiselling, flame cutting or arc gouging, and then the cavity is filled by fusion welding.

3. Exposures in the Workplace

3.1 Introduction

A wide variety of occupational health hazards is present in iron and steel foundries: airborne crystalline silica is virtually ubiquitous in foundries that use quartz sand for moulding and coremaking; metallic fumes are present during melting, pouring, welding and flame-cutting processes; and metal dusts are associated with abrasive grinding operations. The cupola and casting operations may emit carbon monoxide into the working environment. Phenol, formaldehyde, furfuryl alcohol, isocyanates and amines are used as ingredients of organic binders in mould and core sands. Furthermore, several carbonaceous materials are in contact with molten metal during pouring and thus various pyrolysis products, including polynuclear aromatic compounds, are formed. These exposures may occur simultaneously with physical hazards, including noise, vibration, heat and radiation.

The main airborne contaminants to which workers may be exposed are given in Table 4.

Table 4. Airborne substances (and classes of substances) found in iron and steel foundries[a]

Material	Principal uses or sources of emission
Common airborne contaminants	
Amines, aliphatic and aromatic* (e.g., hexamethylenetetramine triethylamine; dimethylethylamine, aniline*)	Urethane binders, amine gassing of urethane resins, thermal decomposition of urea, urethane or shell binders
Ammonia	Thermal decomposition of hexamethylenetetramine in shell moulding, decomposition of urea or urethane binders
Bentonite	Foundry sand, refractory materials
Carbon	Coal powder, graphite and soot in foundry sand, coke in cupola melting, core and mould coatings, constituent of ferrous alloys, electrodes in arc melting and gouging
Carbon dioxide	Combustion of carbonaceous materials in foundry sand, cupola melting, fuel combustion in furnaces, ovens, heaters and engines, carbon dioxide gassing of silicate binders, inert gas welding

Material	Principal uses or sources of emission
Carbon monoxide	Combustion of carbonaceous materials in foundry sand, cupola melting, fuel combustion in furnaces, ovens, heaters and engines, flame cutting and welding
Chromite*	Foundry sand, refractory materials
Chromium and chromium oxides*	Steel alloys, melting, pouring, cutting, grinding and welding operations
Chlorinated hydrocarbons* (e.g., 1,1,1-trichloroethane*)	Solvents
Cristobalite	Refractory materials, high-temperature transformation of silicon dioxide
Fluorides*	Melting, slagging and welding
Formaldehyde*	Urea, phenol and furan resins, thermal decomposition of organic materials in core baking and casting
Furfuryl alcohol	Furan resins
Hydrocarbons, aliphatic and aromatic (e.g., benzene, toluene, xylene, naphthalene)	Solvents for binders and paints, pattern resins and glues, core and mould dressings, metal primers, petroleum fuels, thermal decomposition of organic materials in foundry sand
Hydrogen sulphide	Water quenching of furnace slag, thermal decomposition of sulphur compounds in foundry sand
Iron and iron oxides*	Ferrous alloys, melting, pouring, cutting, grinding and welding
Isocyanates* (e.g., 4,4′-methylenediphenyl diisocyanate*)	Urethane resins, thermal decomposition of urethane binders in foundry sands
Lead and lead oxides*	Scrap melting, spray painting operations
Magnesium and magnesium oxide	Inoculation process in production of nodular iron
Manganese and manganese oxides	Ferrous alloys, melting, pouring, cutting, grinding and welding operations
Nickel and nickel oxides*	Steel alloys, melting, pouring, cutting, grinding and welding operations
Nitrogen oxides	Thermal decomposition of urea or urethane binders in foundry sand, flame cutting and welding, internal combustion engines
Olivine	Foundry sand, refractory materials
Phenols (e.g., cresol, phenol, xylenol)	Phenolic binders, thermal decomposition of organic materials in foundry sand
Polynuclear aromatic hydrocarbons*	Coal-tar pitch, thermal decomposition of carbonaceous materials in foundry sand, fuel combustion in furnaces, ovens, heaters and engines
Silica, quartz	Foundry sand, refractory materials, sand blasting
Sulphur dioxide	Combustion of sulphurous fuels, sulphur-dioxide gassing and decomposition of furan resins
Tridymite	Refractory materials, high-temperature, phase transformation of quartz
Vanadium and vanadium oxides	Steel alloying
Zinc and zinc oxides	Scrap melting
Zircon	Foundry sand, refractory materials
Other airborne contaminants	
Acrolein*	Thermal decomposition of vegetable oils in core baking and casting
Alcohols, aliphatic (e.g., isopropanol)	Solvents for binders and paints, carriers for core and mould dressings, components of urethane resins
Asbestos*	Thermal or electrical insulation in furnaces and ovens; coverings, troughs and clothing in pouring areas
Cadmium and cadmium oxide*	Scrap melting
Calcium carbide, calcium carbonate, calcium silicide, calcium oxide	Melting, alloying and slagging
Carbon disulphide	Decomposition of furan resins with sulphonic acid catalysts
Carbonyl disulphide	Decomposition of furan resins with sulphonic acid catalysts
Copper and copper oxides	Scrap melting, arc gouging with coated carbon electrodes
Cyanides (e.g., hydrogen cyanide)	Thermal decomposition of urea or urethane binders, heat treatment of special castings

Material	Principal uses or sources of emission
Esters (e.g. glycerol diacetate, butyl acetate)	Ester-silicate process, foundry solvents
Ethyl silicate	Silicate binders
Ferrochromium, ferromanganese, ferromolybdenum, ferrosilicon, ferrovanadium	Melting and alloying
Methylethylketone peroxide	Sulphur-dioxide gassing process
Nitrogen heterocyclics (e.g., pyridine)	Coal-tar pitch, thermal decomposition of carbonaceous materials in foundry sand
Nitrosamines* (e.g., *N*-nitrosodimethylamine*, *N*-nitrosodiethylamine*)	Reaction of nitrogen oxides with amines in foundry sand
Oxygen heterocyclics (e.g., furan, methylfuran)	Furan resins
Ozone	Inert gas welding
Phosphine	Reaction of water with phosphides in ferroalloys, decomposition of furan binder, furan resins catalysed with phosphoric acid
Phosphoric acid	Catalyst for furan resins
Radon	Zircon sands
Sulphonic acids (e.g., toluene/sulphonic acid)	Catalyst for furan resins
Sulphur heterocyclics (e.g., thiophene)	Decomposition of furan resins
Talc	Core and mould dressings

[a]This list includes chemicals (or classes of chemicals) used in or formed in iron and steel founding operations, and the processes during which they are used or formed or during which exposures are most likely to occur. It was compiled from information collected during the preparation of the monograph and cannot pretend to be exhaustive.

*Chemicals (and classes of chemicals) indicated by an asterisk have been considered previously in the *IARC Monographs* series and are listed in the Appendix together with the evaluation of carcinogenicity.

3.2 Silica and other mineral dusts

Nearly all production workers in iron and steel foundries are exposed to silica dust and other mineral constituents of foundry sand. Numerous studies on dust exposure have been carried out; however, only the major studies relevant to current work conditions are referred to here.

In 1968-1971, a nationwide survey of exposure to silica was conducted in Sweden. The results are summarized in Table 5 for iron foundries and in Table 6 for steel foundries. The average exposure of the workers was 19.5 mg/m^3 and 10.6 mg/m^3, respectively. Respirable dust contained 12-13% quartz when silica sand was used and about 5% when olivine sand only was used in moulding. In the latter case, the source of quartz dust was mainly the blackings and other siliceous materials used in steel foundries. In 1974, about two-thirds of Swedish steel castings were cast in olivine sand. The heaviest exposure to dust in the iron foundries was during furnace and ladle repair. In sand preparation in iron foundries, exposure to total dust was 16.5, 19.6 and 16.2 mg/m^3 in manual, mechanized and automated operations, respectively. Manual methods such as shovelling, carried out in the proper way, created less dust than the mechanized systems. However, the number of exposed workers is decreased in automated sand plants in which the tasks of the operator are reduced to supervision and maintenance. It was evident from the study that the dust exposure of individual workers was most severe in the automated plants. Moulding sands often contain 50-98% quartz and have a low content of feldspar. Exposure to dust from sandslingers and moulding machines fluctuates widely according to the moisture content of the sand used. In a steel foundry with a wetted floor, the dust concentration was 0.8 mg/m³, while in two foundries with dry waste sand around the moulds, dust concentrations were 34 and 58 mg/m³, respectively. Dust evolved during the use of a sandslinger was usually about 5 mg/m³. At coremaking areas in iron foundries, dust concentrations averaged 4.4 mg/m³ in stationary samples and 6.2 mg/m³ in personal sampling. In a series of measurements in 74 iron foundries, average dust exposures of fettlers grinding with portable and stationary machines were 20.8 mg/m³ and 17.2 mg/m³, respectively. The fettling method that gave the highest exposure was unshielded sand blasting with a jet of quartz particles; in small foundries, an average of 126.9 mg/m³ was recorded. In many cases blasting was done without safety equipment. The average exposure to dust during the repair of furnace and ladle linings in iron foundries was 68.3 mg/m³. The dust contained an average of 22% quartz and 2% cristobalite. In extreme cases, the quartz content of lining

Table 5. Dust exposure in Swedish iron foundries in 1968-1971[a]

Task	Mean exposure to total dust (mg/m³)	Particles $< 5\ \mu m$ (% by weight)	Quartz in respirable fraction (%)
Work with furnaces and ladles	68.3	28	22
Preparing sand	16.2	30	7
Moulding	8.4	33	10
Coremaking	6.2	33	10
General foundry work	13.2	27	16
Shaking-out moulds	16.3	23	9
Shaking-out cores	13.7	23	11
Cleaning castings	17.2	26	13
Cleaning premises	28.4	26	11
Odd jobs	13.6	30	7
Mean of all samples	19.5	27	12

[a]From Gerhardsson (1976); 821 personal samples were collected in 74 foundries.

Table 6. Dust exposure in Swedish steel foundries in 1968-1971[a]

Task	Mean exposure to total dust (mg/m³)	Particles $< 5\ \mu m$ (% by weight)	Quartz in respirable fraction (%)
Preparing sand			
quartz sand	6.4	36	10
olivine sand	30.3	25	5
Moulding			
quartz sand	4.2	31	14
olivine sand	4.9	24	5
Shake-out			
quartz sand	7.6	29	17
olivine sand	8.3	29	6
Cleaning			
quartz sand	10.6	23	11
olivine sand	10.8	15	5
Mean of all samples			
quartz sand	8.8	26	12
olivine sand	12.4	21	5

[a]From Gerhardsson (1976); 92 personal samples were collected in 13 foundries.

dust ranged from 75-90%. In all other operations the concentration of quartz in respirable dust varied from 7-16% in iron foundries and from 5-17% in steel foundries (Gerhardsson *et al.*, 1974; Gerhardsson, 1976).

Similar results were found in a Finnish study of 51 iron and nine steel foundries (Siltanen *et al.*, 1976). The gravimetric concentrations of total dust and respirable quartz are shown in Table 7. Sampling was undertaken in the working environments of 3794 foundry workers (2241 in iron foundries and 1553 in steel foundries). About 30% of them were moulders and coremakers, 25% were fettlers, 15% were melters and pourers, and the rest were crane drivers, plant cleaners, maintenance workers and others. About 70% of the employees engaged in sand preparation, shake-out and fettling were exposed to a respirable quartz concentration exceeding the Finnish exposure standard of 0.2 mg/m^3 (National Finnish Board of Occupational Safety & Health, 1981), and 92% of the samples collected during cupola, furnace and ladle repair operations exceeded this value in steel foundries. The content of respirable dust in the samples was about 30-60% (containing 7-13% quartz) and was similar for most of the foundry operations. Cristobalite or tridymite was found in only a few samples taken from melting departments. Olivine or chromite sand was used in some of the steel foundries, and the content of quartz in respirable dust was found to be lower in these foundries: olivine sand, 1.5% quartz in respirable dust; chromite sand, 2.7%; sea-bottom sand, 5.8-7.8%; and crushed quartzite, 15.1%.

The dust concentrations of different work phases were further analysed with respect to foundry size (Koponen *et al.*, 1976). Dust exposure increased in sand preparation (from 13 to 33 mg/m^3) and melting (from 7 to 16 mg/m^3), decreased in shake-out (from 33 to 10 mg/m^3) and fettling (from 33 to 13 mg/m^3) and was similar in other operations (from 4 to 10 mg/m^3) as plant size increased from small foundries with fewer than 25 employees to larger foundries with over 100 employees. The division also roughly corresponded to the degree of mechanization within the foundries. It was concluded that automated processes allow efficient use of ventilation for some operations. However, with automation, workers may be confined to the same work phase; therefore, in evaluating long-term exposure on the basis of working time

Table 7. Mean concentrations (mg/m^3) of total dust and respirable quartz (particle size, <5 μm) in Finnish iron and steel foundries in 1972-1974[a]

	Iron foundries[b]		Steel foundries[c]	
	Total dust	Respirable quartz	Total dust	Respirable quartz
Sand preparation	21	0.78	13	0.55
Moulding	8.6	0.31	5.8	0.27
Coremaking	5.2	0.24	4.6	0.24
Melting	8.4	0.35	5.0	0.19
Furnace repair	34	2.25	27	5.26
Pouring	8.5	0.19	8.2	0.24
Shake-out	19	0.53	8.6	0.42
Cleaning	18	0.74	21	0.45
Other	9.7	0.41	9.7	0.41

[a]From Siltanen *et al.* (1976)

[b]51 Iron foundries; 1073 samples, 1140 area samples

[c]9 Steel foundries; 342 personal samples, 392 area samples

and dust concentration, the extreme cases were found in larger foundries. In small foundries, workers often have a cycle of various consecutive tasks.

Recent surveys of Italian, Dutch and US foundries also report substantial exposures to silica in foundries (Meyer, 1973; Capodaglio *et al.*, 1976; Zimmerman & Barry, 1976; Bolt, Beranek & Newman, Inc., 1978, as cited in Oudiz *et al.*, 1983; Oudiz *et al.*, 1983).

In an electron microscopic study of nine sands used in foundries, chrysotile asbestos fibres were found in both olivine and chromite sands. These fibres typically had a diameter of less than 0.3 μm; the aspect ratio was 10:1 or greater. Amphibole fibres were also detected in the chromite sand and were somewhat thicker (Stettler *et al.*, 1981). The total concentration of airborne fibres ranged from 0.01 to 0.33 fibres/cm^3 in dust samples collected from five foundries, and the concentration of asbestos ranged from 0.0001 to 0.0033 fibres/cm^3 (Gullickson & Doninger, 1980). In general, the use of asbestos materials has been limited to asbestos textiles in fireproof gloves and aprons or asbestos boards for insulation in melting departments.

Radioactive dust may be inhaled during the use of zircon sand. Its natural radioactivity (from radium 226) was measured as 90 pCi/g (3.33 Bq/g). Radon emanation from zircon sand was estimated to be 2 pCi/m^2 (0.074 Bq/m^2) (Boothe *et al.*, 1980). The main concern regarding radioactivity has been the disposal of waste moulds and sands.

3.3 Metal fumes and dusts

Metal fumes are formed by the evaporation, condensation and oxidation of metals in air. Furnace tenders, melters, casters, ladlemen, pourers and crane drivers are exposed to fumes from molten metal; fettlers are exposed to metal fumes and dusts from grinding, welding and flame-cutting operations.

In 1973-1974, a survey of metal fumes was made in 15 iron foundries and 10 steel foundries in Finland (Tossavainen, 1976). Table 8 gives the average composition of various melting fumes, and Table 9 lists the exposures of furnacemen and casters in foundries with cupola, electric induction or arc melting. The composition of fumes varied with the kind of metal, the quality of scrap and the type of melting process. Altogether, 46 elements were detected by spark-source mass spectrometry. The major compounds were iron oxides (Fe_2O_3 and Fe_3O_4), manganese oxide and zinc oxide. High concentrations of iron, lead and zinc were occasionally found in the workplace air at iron and steel melting areas. The ferroalloys and fluxes added into the charge as well as the constituents of moulding sand contributed to the metal concentration in airborne dusts in melting departments.

Mosher (1980) analysed airborne dust samples taken in 22 US foundries which, at the time of sampling, poured alloys containing less than 2% of chromium and/or nickel. The ferrous foundries melted steel, grey or ductile iron. At least two casting plants were included in each category, and three job titles or locations were chosen for the study: furnace tending, transport and pouring of molten metal and cleaning-room operations, excluding any welding or flame cutting. The results of 94 breathing-zone samples are shown in Table 10. Chromium (0.3-114 μg/m^3) and nickel (0.4-190 μg/m^3) were found in all segments of the industry owing to melting, casting and other operations.

Table 8. Composition of metal fumes by furnace type; average concentration of elements in samples collected in the work areas of furnacemen[a]

Element	Average concentration (%)			
	Steel foundry, electric-induction furnace	Steel foundry, electric-arc furnace	Iron foundry, electric-induction furnace	Iron foundry, cupola
Antimony	<0.1	<0.1	<0.1	<0.1
Barium	<0.5	<0.5	<0.5	<0.5
Cadmium	0.006	0.019	0.004	0.002
Calcium[b]	3.5	11.2	3.0	1.4
Chromium	1.4	0.30	0.19	0.029
Chromium[b]	0.13	0.082	0.043	0.013
Cobalt	<0.02	<0.02	<0.02	<0.02
Copper	0.097	0.10		
Iron	10.2	15.5	20.5	11.3
Lead	1.2	1.7	0.78	0.70
Manganese	1.7	2.6	2.1	0.50
Molybdenum	<0.03	0.15	<0.03	<0.03
Nickel	0.15	0.045	0.038	0.013
Phosphorus	0.29	0.52	0.18	0.33
Potassium	1.0	1.7	0.7	2.0
Silicon	20	15	20	30
Silver	<0.01	<0.01	<0.01	<0.01
Strontium	0.01	0.02	<0.01	<0.01
Sulphur	0.11	0.22	0.57	0.63
Tin	<0.05	<0.05	<0.05	<0.05
Titanium	0.098	0.060	0.092	0.25
Zinc	2.0	7.8	4.1	0.84
Zirconium	2.1	0.57	0.35	0.11

[a]From Tossavainen (1976); 232 area samples were collected in 28 foundries. [b]Acid-soluble

Table 9. Metal concentrations and total dust levels in the air during melting and casting[a]

Exposure	Average concentration ($\mu g/m^3$)			
	Iron foundry, cupola	Iron foundry, electric-induction furnace	Steel foundry, electric-induction furnace	Steel foundry, electric-arc furnace
Cadmium	<2	<2	<2	<2
Calcium[b]	140	230	150	710
Chromium[b]	4	3	5	6
Cobalt	<5	<5	<5	<5
Copper	9	8	8	6
Iron	860	1 590	1 050	620
Lead	15	41	50	49
Manganese	32	130	80	190
Nickel	<5	<5	11	2
Zinc	47	220	110	220
Total dust concentration	10 000	8 600	5 100	6 100

[a]From Tossavainen (1976); 112 personal samples were collected in 27 foundries. [b]Acid-soluble

Table 10. Exposure to chromium and nickel in foundries casting alloys with low or trace levels of chromium (Cr) and nickel (Ni)[a]

Operation	Average concentration (μg/m³)					
	Grey iron		Ductile iron		Steel	
	Cr	Ni	Cr	Ni	Cr	Ni
Furnace tending	3.5	0.4	0.5	0.6	7.1	1.6
Molten metal handling	0.8	0.7	0.3	0.5	5.4	2.1
Fettling room	8.0	6.8	0.9	1.7	113.8	190.7

[a]Adapted from Mosher (1980); 94 personal samples were collected.

Lead concentrations were measured in several grey-iron foundries in the USA that used crushed engine blocks in their charge (Pompeii, 1983). Personal sampling showed lead exposure of the cupola tender to be 80 μg/m³; ladle operator, 110 μg/m³, iron pourer, 60 μg/m³; and cupola charger, 50 μg/m³. In electric-induction melting, the lead exposure was even higher: furnace operator, 1440 μg/m³, ladle operator, 220 μg/m³; iron pourer, 330 μg/m³; and furnace charger, 1140 μg/m³. The source of lead was identified as residues of leaded gasoline in the scrap engines.

Abundant metal fumes are released in the production of ductile iron by the inoculation process that involves the use of Fe-Si-Mg-Mischmetal[1] alloy to 'transform' carbon into nodular form. The concentration of metals (detected by neutron activation analysis) in 1971 in the workplace air in one Belgian foundry varied as follows: iron, 112-9700 μg/m³, magnesium, 32-3400 μg/m³; manganese, 2-60 μg/m³; chromium, 0.4-4.6 μg/m³; nickel, 0.2-10 μg/m³; and zinc 1.7-60 μg/m³. Concentrations of 27 other elements were also reported. Blood samples, while indicating some exposure to lead, were within the limits of analytical errors (Vanhoorne *et al.*, 1972).

In a study of 10 exposed iron foundry workers, the total burden of iron in the lung was estimated by magnetic measurements (Kalliomäki *et al.*, 1979). The amount varied from 30-600 mg among men who had worked for several years as casters, moulders, coremakers or fettlers. In the same study, airborne dust particles examined by electron microscopy were found to be spherical and to form agglomerates and short chains. The single particles averaged about 0.1 μm in diameter. In fettling operations, the airborne particles were irregularly shaped and about 1 μm in size.

Grinding, welding, flame cutting and other fettling operations in iron and steel foundries are similar to those in other metal industries. Because these processes are not especially characteristic of foundry work, although significantly contributing to the overall exposure of the workers, the relevant studies are discussed here only briefly.

The metal concentrations (mg/m³) in 1972-1974 in the breathing zone of welders and flame cutters in six steel foundries were as follows: in manual arc welding (11 personal samples), iron, 14.0; manganese, 1.38; chromium, 0.21; nickel, 0.09; copper, 0.14; lead, 0.10; in flame

[1]Mischmetal contains: 50-55% cerium, 4.5-5.5% praseodymium, 15-17% neodymium, 0.1% samarium and europium; the balance is made up of lanthanum and other rare earth metals.

cutting (22 personal samples), iron, 12.9; manganese, 0.44; chromium, 0.10; nickel, 0.42; copper, 0.05; lead, 0.11. In this study, the copper resulted from the coating of the carbon electrodes. The source of lead was other fettling operations in the vicinity (Virtamo *et al.*, 1975).

Exposure to chromium and nickel among welders working with stainless-steel may be severe. Biological monitoring of urinary chromium showed levels of 11-152 μg/g creatinine in six such welders who were exposed daily to hexavalent chromium concentrations of 0.001-0.52 mg/m^3 in air. Most of the airborne chromium was water soluble and hexavalent (Tola *et al.*, 1977; Tossavainen *et al.*, 1980). A urinary chromium concentration of 40-50 μg/l measured immediately after the work shift corresponds to a level of soluble hexavalent chromium in air of 0.05 mg/m^3 (Gylseth *et al.*, 1977). In a study that included 86 test sites in 18 machine shops, in arc welding of stainless-steel the concentration of chromium in air averaged, with and without local exhaust equipment, 0.06 and 0.2 mg/m^3 as CrO_3, respectively. The nickel concentration was usually less than 0.1 mg/m^3 (Ulfvarson, 1981).

3.4 Carbon monoxide

Carbon monoxide (CO) is given off whenever there is incomplete combustion of carbonaceous material. In the foundry environment, it is mainly produced by cupola melting and casting operations. Other sources are oil burners used for charge and ladle preheating, furnaces for annealing and carburizing, transport equipment powered by internal combustion engines and fettling operations such as welding and flame cutting.

Workers in the furnace area may be exposed to gases leaking from the charging and bottom levels of the cupolas. About 20-30% of the gas emmitted is CO. Typical CO concentrations in the cupola locations were reported as follows (ml/m^3): charging deck, 10-600; iron and slag spout, 15-200; and floor level, 10-300 (Tubich, 1981). Around the cupolas in six iron foundries, a concentration of 50 ml/m^3 was exceeded in 76% of the samples, the average being 240 ml/m^3. Electric furnaces do not evolve much gaseous emission, except during oxygen lancing when the effluents may contain in excess of 80% CO. However, spreading of flue gases was limited when exhaust ventilation was applied directly to the furnace shell. In steel foundries, the CO concentrations were less than 20 ml/m^3 in the areas around the electric furnaces (Virtamo & Tossavainen, 1976a).

Most ferrous castings are produced in green-sand moulds. During pouring, the pyrolysis of carbonaceous material produces gaseous emissions such as carbon monoxide, carbon dioxide, hydrogen, methane and other hydrocarbons. Table 11 lists the composition of effluent gases from a moulding sand containing 6% bentonite, 1% cereal, 3% seacoal and 3.8% water. The experimental design in the laboratory excluded any dilution of the gaseous constituents. The CO concentration averaged about 20%; hydrogen, 58%; carbon dioxide, 6%; and methane, 10%. Ammonia, sulphur dioxide and hydrogen cyanide were found in minor amounts. CO was produced mainly by the reaction of carbon with water to give CO and hydrogen, and experiments have shown that most of the oxygen required for CO synthesis is supplied by sand moisture and not by ambient air. The gas volume increased linearly with the casting weight, i.e., as a function of sand-to-metal ratio. The volume of emitted CO also increased as the content of various carbonaceous additives was increased (Gwin *et al.*, 1976; Scott *et al.*, 1976). Coal-powder replacements may also be used in green-sand moulds, which reduces CO emissions during casting (Hespers, 1972). [However, other contaminants may arise from the pyrolysis of these substitutes.]

Table 11. Composition of gaseous effluents from green sand[a]

Compound	Average concentration
Hydrogen	58 %
Carbon monoxide	20 %
Carbon dioxide	6 %
Methane	10 %
Ethane	0.6 %
Ethylene	1.5 %
Acetylene	0.3 %
Ammonia	4 μg/l
Hydrogen cyanide	90 μg/l
Sulphur dioxide	20 μg/l
Hydrogen sulphide	20 μg/l
Formaldehyde	6 μg/l
(Balance: nitrogen, oxygen and water)	

[a]From Gwin *et al.* (1976)

The workroom concentration of CO rises rapidly during the first few minutes after pouring and then gradually diminishes. A second peak in CO emission occurs during shake-out when hot sand comes into contact with the ambient air (Scott *et al.*, 1976). Near a mechanized shell-moulding line, exposure of workers to CO ranged from 10-50 ml/m^3; peak concentrations reached 600-700 ml/m^3 in a number of plants where resin sands were used in stationary pouring operations. Measurements taken at the breathing zone of operators making shell cores were (ml/m^3): CO, 10-60; ammonia, 7-25; aldehydes, 2-4; and phenols, <1 (Tubich *et al.*, 1960). CO levels of 25-600 ml/m^3 during the pouring of green-sand moulds, 25-400 ml/m^3 during the pouring of resin-bonded moulds and 10-75 ml/m^3 during shell-core moulding were also reported (Tubich, 1981).

In the Finnish study of 52 iron foundries, the concentration of CO averaged 110 ml/m^3 in casting areas during and shortly after pouring. The level of 50 ml/m^3 was exceeded in 72% of 909 short-term samples. In personal sampling of iron casters, the average CO exposure was 85 ml/m^3, with 67% of the two-hour samples exceeding 50 ml/m^3. In steel casting areas, the mean CO concentration was 18 ml/m^3, which was notably lower than that in iron foundries. The blood carboxyhaemoglobin levels of 145 workers (cupola tenders, casters and crane operators) from six iron foundries after the completion of a work shift exceeded 6% in 26% of non-smokers and in 71% of smokers (Virtamo & Tossavainen, 1976a).

3.5 Binder compounds

Organic binder materials for cores and moulds include furan, phenol-formaldehyde, urea-formaldehyde and urethane resins as well as oleoresinous oils. The ingredients may volatilize into the workplace air during mixing, blowing, ramming, drying or baking operations. Curing reactions and thermal decomposition create additional compounds which are driven off in pouring and shake-out. When organic binders are subjected to high temperatures, thermo-oxidative pyrolysis produces gases and smoke aerosols. Only a few of the components have been identified: aliphatic components include methane, ethane, ethylene, acetylene, and lesser amounts of compounds of high molecular weight; and aromatic compounds include benzene, toluene, xylenes, naphthalenes and a variety of polynuclear aromatic compounds, present in lower concentrations. Nitrogen compounds such as ammonia, cyanides and amines may be produced from nitrogen in the urea, ammonium salts and hexamethylenetetramine used as binder chemicals. Urethane resins may emit free isocyanates under moulding and pouring conditions. No-bake catalysts, based on arylsulphonic acids, may produce sulphur dioxide and hydrogen sulphide by thermal processes. If phosphoric acid is used as a catalyst, phosphine can be formed in the extremely reducing atmosphere of the hot emissions. In air, phosphine rapidly oxidizes to phosphorus oxide. Furan binders contain free furfuryl alcohol which can volatilize during mixing, moulding or coremaking. Similarly, furan and phenolic resins may emit formaldehyde, phenol and other derivatives by volatilization or thermal decomposition. Core oils and alkyd isocyanate resins are partly composed of natural drying oils, and heating of these materials gives rise to acrolein, various aldehydes, ketones, acids and esters as well as aliphatic hydrocarbons. When organic solvents are used in sand binders, the vapours may add to the exposure of workers (Toeniskoetter & Schafer, 1977).

Laboratory experiments have indicated the presence of numerous compounds in mould effluents of iron casting. Table 12 shows the results of tests of 12 binder systems, extrapolated to provide an indication of the actual concentrations to be expected at the workplace. The concentrations of contaminant measured from each binder system have been normalized to a carbon monoxide level of 50 ml/m^3 (Scott *et al.*, 1977). The data on nitrogen-containing compounds in mould emissions were reported separately (Table 13). The measured concentrations of nitrogen oxides, ammonia and hydrogen cyanide in mould effluents do not represent workplace conditions but these can be extrapolated by applying a dilution factor of about 30. Low concentrations of aromatic and aliphatic nitrogen-containing compounds such as aniline, morpholine, dimethylformamide, dimethylacetamide, dimethylaniline, triethylamine, *ortho*-toluidine, 2,4-xylidine, *ortho*-anisidine, *para*-anisidine and mesidine were also identified in the effluent by gas chromatography (Emory *et al.*, 1978).

In a laboratory study of binder materials, the following were measured (ml/m^3) in undiluted pouring effluents: carbon monoxide, 215-290; hydrogen cyanide, 1-2; ammonia, $<$1-17; phenol, $<$1-7; and 4,4′-methylenediphenyl diisocyanate (MDI), $<$0.001-0.2 (Toeniskoetter & Schafer, 1977). Ammonia and hydrogen cyanide emissions were associated with urea-modified binders and with shell moulding; airborne MDI was detected in the tests of urethane resins.

In the 1970s, workplace measurements were made for MDI and methylene bisaniline (MBA) in 29 iron and steel foundries. MBA was determined simultaneously because urethane resins emit aromatic amines along with free isocyanates during pouring, cooling and shake-out. Of a total of about 200 samples, 18 had MDI levels in excess of 0.005 ml/m^3 and two had MBA levels in excess of 0.1 ml/m^3. Exposure to phenol and formaldehyde was also measured at the mixing stations in 18 foundries utilizing phenolic urethane no-bake resins. The concentration of phenol ranged from 1-5 ml/m^3 in five of the 90 samples collected; formaldehyde was seldom in excess of 3 ml/m^3. When triethylamine or dimethylethylamine was used as a gassing agent in the urethane process, about 12% of the amine samples showed levels of $>$25 ml/m^3 with an additional 25% of the samples in the range of 10-25 ml/m^3. High amine exposure at a phenolic urethane core-making station was usually the result of leaking fittings, inadequate core-box seals or excessive use of the amine catalyst (Toeniskoetter, 1981). Exposure to solvent vapours was low: 0.2-4 ml/m^3 for aromatic hydrocarbons as measured by personal sampling (Toeniskoetter & Schafer, 1977). Rosenberg and Pfäffli (1982) measured free MDI concentrations in a foundry by high-performance liquid chromatographic and colorimetric methods; they found levels of 0.002-0.009 mg/m^3 in core-making and 0.003-0.040 mg/m^3 in casting.

Table 12. Constituent concentrations of mould effluents normalized to a carbon monoxide level of 50 ml/m³[a]

Effluent[b]	Binder system											
	Green sand	Dry sand	Sodium silicate ester	Core oil	Alkyd iso-cyanate	Phenolic urethane	Phenolic no-bake	Low N_2 furan-H_3PO_4[c]	Medium furan-TSA[d]	Furan hot-box	Phenolic hot-box	Shell (phenolic)
(Dilution factor)	(32.0)	(42.2)	(28.4)	(61.0)	(21.8)	(14.0)	(19.0)	(42.0)	(20.2)	(30.2)	(28.8)	(58.4)
Carbon monoxide† (30 min)	50	50	50	50	50	50	50	50	50	50	50	50
Carbon dioxide† (30 min)	100	110	40	50	150	130	50	35	145	80	85	75
Total hydrocarbon† (30 min)	27	8	13	15	68	58	20	7	32	10	9	5
Carbonyl sulphide† (30 min)	-	0.11	<0.01	<0.01	0.14	0.33	0.14	0.02	0.42	0.23	0.06	0.01
Hydrogen sulphide† (30 min)	-	0.69	0.07	0.03	0.08	0.01	0.95	0.19	1.04	0.04	0.01	0.01
Sulphur dioxide† (30 min)	-	0.97	0.03	0.02	0.04	0.02	2.00	0.29	-	0.02	0.02	0.02
Sulphur dioxide	0.38	1.21	0.09	0.04	0.05	0.10	16.32	0.36	5.94	0.09	0.04	0.51
Hydrogen sulphide	1.23	0.69	0.07	0.02	0.01	0.09	1.58	0.24	0.59	0.06	0.01	0.01
Phenols	0.19	0.09	0.10	0.02	0.14	6.36	1.05	0.01	0.12	0.02	0.23	0.36
Benzene	0.91	0.69	0.53	0.80	6.70	8.71	12.11	0.38	5.54	0.56	1.15	0.98
Toluene	0.09	0.19	0.11	0.16	1.93	1.36	0.68	0.07	10.79	0.03	0.21	0.41
meta-Xylene	0.01	0.09	0.04	0.08	3.16	0.71	0.10	1.31	0.30	0.03	0.14	0.08
ortho-Xylene	0.01	0.05	0.03	0.10	4.82	0.21	0.04	0.43	0.05	0.01	0.03	0.01
Naphthalene	0.02	0.02	0.01	0.01	0.05	0.01	0.02	0.02	0.05	0.02	0.02	0.01
Formaldehyde	0.01	0.01	0.06	0.03	0.13	0.04	0.01	0.16	0.08	0.01	0.01	0.01
Acrolein	0.01	0.02	0.01	0.03	0.11	0.05	0.01	0.02	0.02	0.01	0.01	0.01
Total aldehydes (C_2-C_5)	0.01	0.19	0.49	0.26	2.71	0.20	3.32	0.14	20.79	0.13	0.31	0.06
Nitrogen oxides	0.83	0.12	0.01	0.03	0.44	0.07	0.03	0.01	0.38	0.43	0.73	0.15
Hydrogen cyanide	0.18	0.05	0.07	0.03	0.22	1.71	0.03	0.22	0.74	3.64	1.35	1.54
Ammonia	0.10	0.31	0.01	0.01	0.05	0.14	0.04	0.02	0.25	20.53	12.50	0.56
Total aromatic amines	0.01	0.07	0.02	0.03	0.03	0.57	0.02	0.05	0.45	3.18	1.46	0.34

[a]From Scott *et al.* (1977)

[b]Concentrations measured in mg/m³ except for those indicated by † which were measured in ml/m³

[c]Phosphoric acid

[d]Toluenesulphonic acid

Table 13. Nitrogen compounds in undiluted mould emissions[a]

Binder system	Concentration (mg/m^3)		
	Nitrogen oxides	Ammonia	Hydrogen cyanide
Alkyd isocyanate	9.7	1.2	24.7
Phenolic urethane	3.1	2.4	24.5
Shell	8.5	33.0	90.0
Phenolic no-bake	0.6	1.0	0.6
Nitrogen-free furan no-bake	0.3	0.5	0.8
Furan no-bake	7.7	5.0	15.0

[a]From Emory *et al.* (1978)

Levels of furfuryl alcohol, formaldehyde, phenol and phosphoric acid were measured in several iron and steel foundries using furan resins. In the air of core-making areas, mean furfuryl alcohol and formaldehyde concentrations were 4.3 ml/m^3 and 2.7 ml/m^3, respectively. Furfuryl alcohol concentration exceeded 5 ml/m^3 in 22% of the samples, and formaldehyde concentration exceeded 2 ml/m^3 in 38%. The highest phenol concentration was 0.35 ml/m^3, and all phosphoric acid samples contained less than 0.1 mg/m^3 (Virtamo & Tossavainen, 1976b).

In another study of furan resins, 75 measurements were made in the mixing areas of 20 foundries (Eftax *et al.*, 1977). All breathing-zone concentrations of furfuryl alcohol were less than 50 ml/m^3, with 80% less than 10 ml/m^3 and 65% less than 5 ml/m^3. Most formaldehyde concentrations were between 0.3-1 ml/m^3. In pouring areas, sulphur dioxide and furfuryl alcohol were found in low concentrations in adequately ventilated areas.

Hexamethylenetetramine is used as an accelerator in shell moulding. The level of its decomposition product, ammonia, has been determined in the breathing zone of machine operators in various studies: 7-25 ml/m^3 (range) (Tubich *et al.*, 1960); 15 mg/m^3 (mean) (Drasche, 1976); and 35 ml/m^3 (mean) (Virtamo *et al.*, 1975). Drasche (1976) also reported the following measurements (mg/m^3) in the shake-out operations of a shell-moulding foundry: phenol, 11; formaldehyde, 0.2; cresols, 11; and benzene, 6.

3.6 Polynuclear aromatic compounds

Polynuclear aromatic compounds (PACs) are formed during thermal decomposition of carbonaceous ingredients in foundry sand. During casting, PACs are synthesized and partly vaporized under the extremely hot and reducing conditions at the mould-metal interface. Thereafter, PACs are adsorbed onto soot, fume or sand particles and spread throughout the workplace during shake-out and other dusty operations. Although the mechanism of PAC formation is complex and variable, the reactions evidently proceed by free-radical pathways. Various radical species containing carbon atoms combine in rapid fashion at the temperature range of 500-800°C. This pyrosynthesis is affected by many variables, such as, the composition of the gaseous atmosphere and the chemical structure of the carbonaceous material. Organic binders, coal powder and other carbonaceous additives are the predominant sources of PACs in iron and steel foundries. In some cases, exhaust gases from engines, furnaces and ovens may increase the exposure of workers to these compounds. The total PAC exposure is composed of numerous compounds that occur concurrently and combine with other air contaminants. As indices of PAC exposure, the airborne concentration of a single compound such as benzo[*a*]pyrene, a sum of specified or unspecified PACs or a benzene-soluble fraction of particulate matter have been used.

(*a*) *Laboratory experiments*

Experimental studies on the formation of PAC from sand mixtures have been reported (Gwin *et al.*, 1976; Novelli & Rinaldi, 1977; Southern Research Institute, 1979; Schimberg *et al.*, 1981a). These studies (described below) employed gas chromatography combined with mass spectrometry as an analytical method; however, the quantitative results are comparable only within the test series in question, owing to experimental differences such as the sand-to-metal weight ratio, interface area and sampling techniques.

In a study in Italy, 13 green-sand additives were pyrolysed at 1000°C in a laboratory oven; measurements of the PACs emitted are shown in Table 14. It should be noted that the emission of the compounds varied about 1000-fold as expressed in mg/kg of additive. The highest value was from coal-tar pitch and the lowest was from vegetable products. The test atmosphere - air or nitrogen - did not affect PAC formation. In further experiments performed in a foundry, 22 kg of grey iron were cast in 50 kg of a sand mixture which contained 89% silica sand, 6.3% bentonite, 3.6% moisture and 0.9% either coal-tar pitch, polymeric resin or maize starch as carbonaceous additive. Each mould thus contained 450 g of the additive to be tested. Emissions were collected in benzene scrubbers. Benzo[*a*]pyrene concentrations measured were 0.230, 0.037 and 0.030 mg/m^3 of gas volume, respectively, for the three additives, corresponding to calculated values of benzo[*a*]pyrene of 32, 6 and 7 mg/m^3 if normalized to a carbon monoxide level of 50 ml/m^3 in ambient air (Novelli & Rinaldi, 1977, 1979).

A total of 50 compounds, identified as PACs, were found in air samples from a pilot-plant foundry and in airborne particulate matter from six iron foundries in Finland (Schimberg *et al.*, 1978, 1980; Schimberg, 1981). These included anthracene, benzo[*b*]chrysene,

Table 14. Formation of four polynuclear aromatic compounds emitted from sand additives in air at 1000°C[a]

Additive	Concentration (mg/kg of additive)			
	Fluoranthene	Pyrene	Benzo-fluoranthene	Benzo[*a*]pyrene
Maize starch	16	26	3	2
Resin	1482	992	1152	22
Coal-tar pitch	2886	3008	3040	3080
Wood flour	179	29	106	6
Pre-gelled starch	71	128	60	35
Petroleum pitch	650	1216	533	336
Polystyrene	2184	1104	3720	1000
Gilsonite	585	960	600	360
Seacoal	187	205	100	134
Seacoal in emulsion	312	416	127	49
Seacoal in oil	234	384	144	14

[a]From Novelli & Rinaldi (1977)

benzo[*b*]-, -[*j*]- and -[*k*]fluoranthenes, benzo[*a*]fluorene, benzo[*ghi*]perylene, benzo[*c*]phenanthrene, benzo[*a*]pyrene, benzo[*e*]pyrene, dibenzanthracenes, fluoranthene, fluorene, *ortho*-phenylenepyrene, perylene, phenanthrene and pyrene and the heterocyclic, benzo[*b*]naphtho-[2,1-*d*]thiophene. The gas chromatograms of all benzene-soluble matter and particulate material retained on a glass-fibre filter indicated that PACs with molecular weights below 200 were not entirely collected and those with molecular weights above 275 could not be completely extracted and analysed by gas chromatography.

Four coal powders, three coal-tar pitches, three polymers and four materials of vegetable origin were compared as sources of benzo[*a*]pyrene emissions in casting tests in a laboratory-scale iron foundry (see Table 15) (Schimberg *et al.*, 1981a; Tossavainen & Schimberg, 1981). Moulding sand was made of silica sand mixed with 5% bentonite, 4% water and 0.5-6% of the carbonaceous additive in question. The volume of emitted gases from the 600-g test mould ranged from 3-10 l and was directly proportional to the percentage of additive. The average gas composition was as follows: carbon monoxide, 19%; carbon dioxide, 6%; nitrogen, 9%; oxygen, 2%; methane, 10%; and hydrogen, 54%. The type and quantity of the particulate pyrolysis products were dependent on the additive. Before casting, the concentrations of benzo[*a*]pyrene in various formulations of moulding sand were <0.05 mg/g of sand, except for the coal-tar pitch formulation in which the concentration was 20-90 μg/g. Most of the benzo[*a*]pyrene present was synthesized during the casting test, in quantities ranging from 1 μg (polymer additives) to 4500 μg (coal products). The emission of benzo[*a*]pyrene from the different formulations decreased in the following order: coal-tar pitch, coal powder, wood flour, peat, tall oil, starch and synthetic polymers. Results from profile analyses of nine PACs were similar within categories of additives - coal-tar pitches, coal powders and products of vegetable origin. The proportion of benzo[*a*]pyrene ranged from 1-17% of the total amount of anthracene, benzo[*k*]fluoranthene, benzo[*e*]pyrene, chrysene, fluoranthene, perylene, phenanthrene and pyrene. Only minor quantities of polynuclear aromatic hydrocarbons were detected from the decomposition of synthetic polymers. Cyclohexane-soluble matter was not further analysed, but the fraction by weight was highest for polymers, coal powders, coal-tar pitch and tall oil. The above tests were all made with new sand mixtures. Similar moulds prepared from old foundry sand emitted an average of 1240 mg of cyclohexane-soluble materials and 44 μg of benzo[*a*]pyrene.

Table 15. Emissions of benzo[*a*]pyrene and total cyclohexane-soluble matter from 600 g of moulding sand in casting experiments[a]

Additive	Proportion in moulding sand (%)	Total cyclohexane-soluble matter (mg)[b]	Benzo[*a*]pyrene (μg)[b]
Coal-tar pitches	0.5-4	460	1630
Coal powders	1-6	850	590
Wood flour	1-4	320	24
Peat	2-4	240	23
Tall-oil pitch	0.5-1.5	1900	10
Starch	1-3	30	3
Petroleum resin	0.5-1.5	980	<1
Polystyrene	0.5-1.5	890	<1
Polyethylene	0.5-1.5	890	<1

[a]From Schimberg *et al.* (1981a)

[b]Mean values calculated by the Working Group

Among over 100 compounds observed in condensable emissions from green-sand moulds were benzo[*a*]pyrene, benzo[*e*]pyrene and perylene (Gwin *et al.*, 1976).

In a study in the USA, grey-iron castings were poured at 1450°C into sand moulds prepared from 12 commercial formulations of selected foundry binders. After pouring, each mould was covered with an exhaust hood and effluent samples were taken. The total relative weights of particulate matter, benzene-soluble materials and benzo[*a*]pyrene are listed in Table 16. The results were calculated by assuming a carbon monoxide dilution of 50 ml/m^3: Levels of benzo[*a*]pyrene of 0.184, 0.322 and 0.026 μg/m^3 corresponded to that level of carbon monoxide for green sand, dry sand and core oil, respectively. These were the highest levels reported, and it should be noted that the above formulations contained seacoal or cereal in addition to resin binders (Southern Research Institute, 1979). (The gaseous products of the resin binders were reported by Scott *et al.*, 1977 and are described on p. 155 of this monograph.)

Table 16. Relative total weights of particulate matter, benzene-soluble materials and benzo[*a*]pyrene recovered from emissions from approximately equal amounts of binder sands[a]

Binder system	Particulates (mg)	Benzene-soluble materials (mg)	Benzo[*a*]pyrene (mg)[b]
Green sand	1652	1021	1.2
Dry sand	430	405	2.5
Sodium silicate ester	17	13	0.01
Core oil	472	355	0.30
Alkyd isocyanate	74	21	0.02
Phenolic urethane	23	7	0.007
Phenolic no-bake	66	61	0.002
Zero-nitrogen furan	359	125	0.016
Medium-nitroge n furan	50	10	0.012
Furan hot box	211	181	0.002
Phenolic hot box	195	75	0.006
Shell phenolic	157	107	0.21

[a]From Southern Research Institute (1979)

[b]Amounts of benzo[*a*]pyrene analysed are very small, and there could be a 20% error in analysis. They were calculated by assuming a carbon monoxide dilution of 50 ml/m^3.

Most of the PAHs and other organic compounds can be extracted from particulate matter with cyclohexane. The soluble portion varied from 16-36% of total particulate emission from furan, urethane, shell and green-sand moulds (Palmer *et al.*, 1981). The major part of the insoluble fraction consisted of carbon particles. The water-insoluble fraction of the four effluents from the furan, urethane, shell and green-sand moulds, was 10-15%, 1-2%, 50% and 20-25%, respectively. The only PACs detected in the water phase were acridine in the furan, green-sand and urethane effluents and phenanthrene in the shell and green-sand samples. Otherwise, the measured PACs were found in the cyclohexane extracts. Phenols were associated entirely with the water-soluble fraction, and their concentrations varied, as shown in Table 17.

Table 17. Concentrations of phenols in water-soluble fractions of total particulate emissions from various types of moulds[a]

Phenol	Concentration (μg/g)			
	Furan	Urethane	Shell	Green-sand
Phenol	2 000	50 000	1 600	1 000
Pentachlorophenol	3	<2	<2	<2
4-Nitrophenol	48	420	1 800	<2
2-Nitrophenol	<2	<2	2 300	<2
2,4-Dimethylphenol	<2	21	60	<2

[a]From Palmer *et al.* (1981)

(*b*) *Workplace measurements*

The first measurements of benzo[*a*]pyrene in the foundry environment were reported by Zdražil and Pícha (1963, 1965). Before casting, airborne levels of benzo[*a*]pyrene were 0.03-0.12 μg/m³ in sand preparation, moulding, shake-out, fettling and sand-blasting operations; after casting, levels were 0.08 μg/m³ in machine moulding, 0.13-0.47 μg/m³ in shake-out and 0.38 μg/m³ in fettling operations. When coal-tar pitch was used in moulding sand, levels were substantially increased.

In 1976-1977, a preliminary survey in a Canadian steel foundry found concentrations of benzo[*a*]pyrene of 0.024-0.295 μg/m³ benzo[*a*]pyrene in addition to other PACs (Gibson *et al.*, 1977). The environmental study was later completed by personal and area sampling in 10 ferrous and non-ferrous foundries. The results for benzo[*a*]pyrene, benzo[*a*]anthracene/chrysene and dibenz[*a,h*]anthracene are shown in Table 18, as classified by occupational category. The study included three iron foundries and three steel foundries in which benzo[*a*]pyrene concentrations averaged 0.94 and 0.43 μg/m³, respectively. The values were much higher than those measured in non-ferrous foundries (Verma *et al.*, 1982).

The concentrations of benzo[*a*]pyrene measured in the workplace air of six Finnish iron foundries during melting, moulding, shake-out and fettling operations are given in Table 19. Two of the foundries used coal-tar pitch and four used coal powder as the moulding-sand additive. Otherwise the process and hygienic conditions in the foundries were similar, as confirmed by measurements of total dust concentrations. Coal-tar pitch produced an average concentration of benzo[*a*]pyrene of 5.1 μg/m³, which was about 60 times higher than the mean of 0.08 μg/m³ measured in foundries using coal powder. In the former category of foundries, a mean level of benzo[*a*]pyrene of 1 μg/m³ was exceeded in moulding, casting, shake-out and transport operations, while in the latter category none of 76 samples was in excess of that level. One exceptionally high concentration of benzo[*a*]pyrene (72 μg/m³) was found close to a defective conveyor in a sand plant in an iron foundry using coal-tar pitch. In the six foundries the proportion of benzo[*a*]pyrene comprised about one third of all PACs measured, i.e., benzo[*k*]fluoranthene, benzo[*e*]pyrene, benzo[*a*]pyrene and perylene (Schimberg *et al.*, 1980). More than 80% of the benzo[*a*]pyrene was associated with dust particles below 5 μm in size (Schimberg, 1981).

Table 18. Mean and maximum contentrations of polynuclear aromatic hydrocarbons in 10 Canadian foundries[a]

Occupation	Number of observations	TSP[b] (mg/m^3)		Total PAH (% of TSP)		B[a]A[c]/chrysene (μg/m^3)		B[a]P[d] (μg/m^3)		DB[a,h]A[e] (μg/m^3)	
		Mean	Maximum	Mean	Maximum	Mean	Maximum	Mean	Maximum	Mean	Maximum
Mixer	8	9.64	34.81	0.02	0.07	0.46	1.27	0.42	2.01	0.28	0.74
Moulder	17	4.36	8.38	0.14	0.76	1.72	9.39	0.67	2.74	0.74	4.14
Pourer	9	4.96	9.96	0.16	0.45	1.37	2.99	0.87	3.44	1.92	12.54
Shake-out	10	6.07	16.48	0.15	0.99	1.80	6.45	1.18	8.15	0.62	3.69
Meltman	12	6.08	16.88	0.10	0.57	1.35	5.08	0.64	5.49	0.49	2.23
Finishing	11	7.64	14.48	0.01	0.06	0.14	0.59	0.09	0.52	0.09	0.61
Coremaker	16	5.38	42.56	0.10	0.69	0.76	5.00	0.17	0.81	0.09	0.81
Craneman	9	3.63	5.78	0.18	0.49	1.69	5.19	0.44	1.64	0.58	3.13
Area sample	10	7.97	52.17	0.52	4.10	2.90	22.10	1.78	12.06	1.29	9.48

[a]From Verma *et al.* (1982)

[b]TSP, total suspended particulate

[c]B[*a*]A, benzo[*a*]anthracene

[d]B[*a*]P, benzo[*a*]pyrene

[e]DB[*a*,*h*]A, dibenz[*a*,*h*]anthracene

Table 19. Concentrations of benzo[*a*]pyrene in workplace air in iron foundries using coal-tar pitch or coal powder as a moulding sand additive[a]

Operation	Mean concentration of benzo[*a*]pyrene (μg/m³)	
	Two foundries using coal-tar pitch	Four foundries using coal powder
Melting	0.2	0.01
Moulding	2.2	0.04
Casting	2.1	0.09
Shake-out	12.6	0.10
Fettling	0.5	0.11
Transport and others	1.4	0.04
Mean concentration of benzo[*a*]pyrene	5.1 μg/m³	0.08 μg/m³
Mean concentration of total dust	10.2 mg/m³	7.8 mg/m³

[a]From Schimberg *et al.* (1980)

4. Biological Data Relevant to the Evaluation of Carcinogenic Risk to Humans

The studies of carcinogenicity and mutagenicity reported in this section include only those on samples of complex mixtures taken from the industry.

4.1 Carcinogenicity studies in animals

No data were available to the Working Group.

4.2 Other relevant biological data

(*a*) *Experimental systems*

No data were available to the Working Group on toxic and reproductive effects.

Tests for genetic and related effects in experimental systems

Extracts of particulate samples collected at several sites in a steel foundry were tested for mutagenicity in *Salmonella typhimurium*. The air samples were collected on glass-fibre filters, and Soxhlet methanol extracts (redissolved in dimethylsulphoxide) were prepared for testing.

Air samples collected in the breathing zone at the pouring-floor level were mutagenic to strain TA98 both in the presence and absence of an Aroclor-induced rat-liver metabolic system (S9). The aqueous, basic, acid and neutral fractions of these samples were all mutagenic to strain TA98 (Kaiser *et al.*, 1980). An additional 16 air samples taken in the breathing zone at the floor level of the same foundry were mutagenic to strain TA98 with and without S9; no mutagenic activity was observed in strain TA100 (Kaiser *et al.*, 1981).

A series of samples were taken from locations chosen to correspond generally with the work areas identified by Gibson *et al.* (1977) in their epidemiological study. The air samples included areas on the overhead crane runways over the core making, moulding, pouring and finishing areas and in the floor breathing zone. The samples were all mutagenic to strain TA98, both with and without S9, and the authors report that an analysis of variance shows the differences between these work sites to be significant [statistics not presented]. Samples taken over the core-making and moulding sites were more mutagenic than samples from the finishing and furnace areas. All of the air samples taken from various sites within the foundry were more mutagenic than the air outside the foundry (Kaiser *et al.*, 1981; Bryant & McCalla, 1982).

Soxhlet cyclohexane extracts of airborne particulates collected on glass-fibre filters from the breathing zone of workers in two iron foundries using coal-tar pitch as an additive in the moulding sand were mutagenic to *S. typhimurium* strains TA98 and TA100 in the presence of a rat-liver metabolic system. In only one of the foundries was the concentration of benzo[*a*]pyrene highly correlated (correlation coefficients 0.78 and 0.87) with the mutagenic activity (Skyttä *et al.*, 1980).

(*b*) *Effects in humans (other than cancer)*

Toxic effects

Silicosis. Rüttner (1954) described pneumoconiosis in 21 autopsies on foundry workers in Zurich; nine were complicated by tuberculosis. The severity of the pneumoconiosis was proportional to length of exposure. Rüttner claimed that, microscopically, the basic lesion of the pneumoconiosis of foundry workers is the anthracosilicotic nodule.

McLaughlin and Harding (1956) collected 149 cases of pneumoconiosis in foundry workers. The first series comprised 64 cases, which were those referred to Pneumoconiosis Medical Panels, and the second series comprised 85 cases, all foundry workers who died in the Sheffield area (UK) between 1949 and 1954 and on whom an autopsy was performed. Silicosis or mixed-dust fibrosis was seen in all cases in the first series and in 89% of those in the second; 43.6% of all cases had tuberculosis in some degree of severity. Of the total of 149 cases, 16 (10.8%) had carcinoma of the bronchus. Occupation within the foundry showed an influence on mortality: there was a serious risk of silicosis and early death in steel fettlers, but steel moulding appeared to be one of the least hazardous occupations in the foundry industry. Sand and shot blasters, in particular, had a high incidence of silicosis and a shortened life span.

McLaughlin (1957) reviewed pneumoconiosis in 2767 foundry workers in the UK. The group was classified into three broad sub-groups: moulding, fettling and others; in all three sub-groups, steel workers showed the highest incidence of abnormal X-ray changes, with the highest incidence among fettlers. In all types of foundries, the highest incidence of X-ray abnormalities was found in the fettling shops.

MacBain *et al.* (1962), in a larger survey on foundries, found some chest abnormality in 307 of 2163 (14.2%) men working in iron foundries; 96 cases were regarded as pneumoconiosis. Of 1462 men who had not previously been employed in coal or other mines, 55 (3.8%) showed pneumoconiosis. [The films were read by only one reader, but random sampling and re-reading at a later date indicated consistency in interpretation.]

Gregory (1970) surveyed a single steel foundry in Sheffield, UK, between the years 1950 and 1960. X-rays, taken in mobile units, were read by two radiologists and classified according to the International Labour Office classification. Of a total of 3708 workers, 877 were employed in the main foundry and fettling and welding shops. Among these, there were 90 cases of pneumoconiosis, considered to comprise 60 with silicosis, 18 with siderosis, six with coal workers' pneumoconiosis and six with mixed dust changes. There were 25 cases of progressive massive fibrosis.

The prevalence of pneumoconiosis, chronic bronchitis and lung function was studied among a sample of 931 men with the longest exposure time who responded to a survey of 15 401 Finnish foundry workers employed during 1950-1972 (Hernberg, 1976; Hernberg *et al.*, 1976). The mean exposure time was 17 years (minimum exposure time, 4.2 years; standard deviation, 9 years). The prevalence of pneumoconiosis was estimated to be 3.8%, increasing to 5.4% for exposure longer than ten years (Kärävä *et al.*, 1976).

Chronic bronchitis. The prevalence of chronic bronchitis among foundry workers in Finland increased with cigarette smoking, being 11% for non-smokers, 24% for ex-smokers and 61% for smokers (Kärävä *et al.*, 1976). Chronic bronchitis was highest in smokers with high dust exposure.

Occupational asthma. Occupational asthma has been described in foundry workers, particularly since the introduction of new binders for sand moulding. For example, Cockcroft *et al.* (1980) described severe asthma in a 50-year-old moulder a few weeks after he had commenced work with a furan binder.

Siderosis. Non-fibrotic pneumoconiosis may develop from the inhalation of metallic dusts and fumes present in foundry environments. Exposure to iron compounds, for example, may cause siderosis, a non-symptomatic lung disease. Siderosis is detectable as radiographic changes (opacities) in the lung. The opacities represent deposits of iron oxide on alveolar, septal and perivascular walls of the lung (Morgan & Seaton, 1975). If fibrogenic dusts (e.g., silica) are inhaled with the iron compounds, a fibrotic lung response may result from this 'mixed dust' exposure.

Dermatitis. Dahlquist (1981) reported contact dermatitis in a woman employed at a cast-iron foundry to mend sand cores with a product containing colophony resin and a small amount of formaldehyde. Patch testing showed contact allergy to these two substances. The dermatitis disappeared completely when the woman was absent from work because of a fracture, and recurred one week after she resumed work.

Mutagenicity in urine

Mutagenic activity in *S. typhimurium* strain TA98 in the presence of a rat-liver activation system was measured in urine samples from eight smoking and four non-smoking workers in an iron foundry. Mutagenicity was found in all samples, and the activity was related neither to the benzo[*a*]pyrene concentration in air nor to the total dust concentration in air (Schimberg *et al.*, 1981b). [The Working Group noted the small number of samples and the lack of non-exposed smoking and non-smoking controls which make this study difficult to interpret.]

In a similar study, it was found that the mutagenic activity in urine from 10 smokers and 10 non-smokers in iron foundries was lower after a four-week holiday than in samples collected at the end of the work shift (Falck, 1982).

4.3 Case reports and epidemiological studies of carcinogenicity in humans

For reviews, see Tola (1980) and Palmer and Scott (1981).

(*a*) *Case reports*

McLaughlin and Harding (1956) reported clinical and pathological observations in 85 cases of iron and steel foundry workers, collected in 1956, irrespective of the suspicion of pneumoconiosis, who died in the Sheffield, UK, area between 1949 and 1954, and on whom autopsy could be obtained, and in 64 cases collected in 1950 among those referred to the Pneumoconiosis Medical Panels with a view to compensation. They found carcinomas of the bronchus in 16/149 (10.8%): 13/85 (15.3%) in the 1950 series and 3/64 (4.7%) in the 1950 series. The authors claimed that these percentages [10.8 and 15.3] 'are notably higher than in the general population of similar age'. [An occupational history was obtained for each case, but the small number of cases prevented any meaningful comparison across specific occupational categories.]

(*b*) *Surveys on mortality statistics* (see Table 20)

All cancer deaths occurring among males over 14 years of age in the city of Sheffield, UK, during the period 1926-1935 were investigated by Turner and Grace (1938). Cause of death, age at death and occupation were obtained from the official data sheets returned to the Medical Officer of Health by the local registrars, and the numbers of workers in various occupational groups in the Sheffield area were supplied by the Registrar General; 3816 cancer deaths in 14 occupational groups were investigated. Expected deaths from different cancer sites within each group were calculated from the age-specific death rates of business and professional men, and clerks. The occupational group 'iron and steel foundry workers, furnace workers, smiths, etc.' showed the greatest excess from the following causes of death: all cancers (observed, 758; expected, 402); cancer of larynx, lung and bronchi and mediastinum (126/54); cancer of the buccal cavity and pharynx (115/33); cancer of the alimentary tract (346/214), with 162 deaths from stomach cancer compared to 81 expected; cancer of the skin (30/4). There was a lower than expected mortality from liver and gall-bladder cancer (23/27), and mortality from cancer of the pancreas was equal to expectations (11/11). The occupational group 'engineers, machinists, cutlers, toolmakers, turners, etc.' among iron and steel works experienced very similar mortality patterns; workers in this group had the second highest mortality for the cancer sites considered. Moreover, they experienced the highest mortality from cancer of the liver and gall-bladder and of the prostate and pancreas. [Some of the cancer mortality excess observed in the study may be accounted for by differences in socioeconomic grouping between the study and comparison groups.]

Swantson (1950) surveyed data from the Registrar General's Decennial Supplement for England and Wales, 1930-1932, with special attention to the mortality experience of iron and steel workers (287 229 males aged 20-65 years). The occupational group 'furnacemen, rollers and skilled assistants' showed excesses of cancers of all sites (ratio of observed to expected deaths x 100 (SMR), 154), of cancer of buccal cavity and pharynx (250), of cancer of the lung (160, based on 16 deaths observed), and of cancer of the oesophagus and stomach (153). The group 'metal moulders and die-casters' had an excess of all cancers (130), lung cancer (193) and cancer of the oesophagus and stomach (125). Among 'iron foundry furnacemen and labourers', the SMR for all cancers was 150; for cancer of the buccal cavity and pharynx, 200; and for oesophageal and stomach cancer, 140; 15 deaths from lung cancer were observed, yielding a ratio of 188. These occupational groups also experienced an excess of deaths from all causes (121, 112, 132, respectively). Expected deaths for each group were calculated on the basis of the age- and cause-specific mortality rates for all males aged 25-64 years, obtained from the census. Mortality from specific cancer sites exhibited a gradient according to socioeconomic grouping, except that the lung cancer distribution appeared to be fairly homogeneous across the groupings.

Morrison (1957) reported data on mortality by occupation from the Registrar General for Scotland, 1956. Only those causes of death for which a minimum of 50 cases were observed were included in the report. Expected deaths within specific occupations were calculated from the age-specific death rates for all males aged 15-64 years. 'Moulders' showed a SMR of 162 for lung cancer, ranking third among the occupational groups considered. 'Foundry labourers, etc.' had a ratio of 161 for lung cancer, which was the fifth highest.

The occupational mortality tables for England and Wales, based on the 1961 census, were reviewed by Adelstein (1972). Mortality from different cancer sites among 'occupied' and retired males aged 15-74 years was classified into 27 occupation orders. Expected deaths within specific occupations were calculated from the age-specific proportion of deaths from a given cause, out of all causes, for all men (standardized proportionate mortality ratio, PMR). Occupation order V ('furnace, forge, foundry and rolling mill workers') showed significantly elevated PMRs for lung cancer (123, third highest among the 27 orders) and for all cancers (110, fourth highest).

Other descriptive data regarding cancer mortality and occupations, based upon the decennial analysis of the Registrar General of England and Wales, are reported by Logan (1982). Expected deaths were obtained from age-specific death rates of all men in the period surveyed. From 1960-1962, the SMR for lung cancer in the occupation order V was 140 (third highest); also elevated and among the highest were the ratios for all cancers (125), and for rectal (133) and pancreatic cancer (131). During the period 1970-1972, the SMRs were further standardized taking into account differences of socioeconomic grouping. Occupation order V had, in this review, the highest excesses for lung cancer (155) and for leukaemia (154), and the second highest for all cancers (135). [It is worth noting that while this occupation order ranked first with regard to mortality from lung cancer, it ranked only fifth according to the smoking score attributed to each order (Office of Population Censuses & Surveys, 1978).]

The Registrar General's Decennial Supplement for England and Wales 1970-1972 (Office of Population Censuses & Surveys, 1978) reported a more detailed analysis by occupation. 'Moulders and coremakers' exhibited a SMR of 130 for all causes and of 184 for lung cancer (based on 155 observed deaths). 'Fettlers' had a ratio of 110 from all causes, of 129 from lung cancer, and of 161 from stomach cancer. Finally, 'metal furnacemen' showed a ratio of 113 for all causes and of 155 for lung cancer.

Mortality tables prepared by the US National Vital Statistics Division for the year 1950 were the basis of the analysis of mortality from lung cancer by occupation carried out by Enterline and McKiever (1963). Only deaths among males aged 20-64 years were considered. Expected

Table 20. Epidemiological studies of iron and steel foundry workers: Mortality statistics

Reference	Source of data	Industry	Occupation	Cause of death	Deaths		SMR[a]	Comments
					Observed	Expected		
Turner & Grace (1938)	Medical Officer of Health, Sheffield, UK, and Registrar General's 1931 census	Iron and steel works	Foundry workers, furnace workers, smiths, etc.	All cancers	758	402	189	All cancer deaths of males over 14 years of age in the city of Sheffield in the period 1926-1935 were surveyed; expected deaths obtained from mortality of professional and business men and clerks
				Respiratory system	126	54	233	
				Buccal cavity and pharynx	115	33	348	
				Alimentary tract	346	214	162	
				Stomach	162	81	200	
				Rectum	62	44	141	
				Liver, gallbladder	23	27	85	
				Skin	30	4	750	
Swantson (1950)	Registrar General's Decennial Supplement, England and Wales 1930-1932	Iron and steel industry	Furnace men, rollers and assistants	All causes			121	For males aged 20-65 years
				All cancers			154	
				Lung	16	10	160	
				Buccal cavity and pharynx			250	
			Metal moulders and die-casters	All causes			112	
				All cancers			130	
				Oesophagus and stomach			125	
				Lung			193	
			Iron foundry furnacemen and labourers	All causes			132	
				All cancers			150	
				Buccal cavity and pharynx			200	
				Oesophagus and stomach			140	
				Lung cancer	15	8	188	
Morrison (1957)	Registrar General for Scotland 1956	Occupations coded according to General Register 1951	Moulders	Malignant neoplasm of respiratory system			162	For males aged 15-64 years; only SMRs based on 50 or more deaths were reported
			Foundry workers	Malignant neoplasm of respiratory system			161	
Adelstein (1972)	Registrar General's Decennial Supplement, England and Wales 1960-1962		Furnace, forge, foundry, rolling-mill workers	All cancers			110*	PMR[b]
				Lung cancer			123*	

Logan (1982)	Registrar General's Decennial Supplements England and Wales 1960-1962 and 1970-1972		Furnace, forge, foundry, rolling-mill workers	All causes (1961)			106	For 1971 standardized by social class
				All causes (1971)			122	
				All cancers (1961)			125	
				All cancers (1971)			135	
				Oesophagus (1961)			127	
				Oesophagus (1971)			120	
				Stomach (1961)			121	
				Stomach (1971)			144	
				Rectum (1961)			133	Second highest of 27 occupation orders
				Rectum (1971)			141	
				Lung (1961)			140	Third highest
				Lung (1971)	515[c]		155	Highest
				Bladder (1961)			129	
				Bladder (1971)			133	
				Leukaemia (1961)			87	
				Leukaemia (1971)			154	Highest
Office of Population Censuses and Surveys (1978)	Registrar General's Decennial Supplement. Occupational Mortality 1970-1972	Foundry workers	Moulders	All causes			130	
				Lung cancer	155		184*	
			Fettlers	All causes			110	
				Lung cancer			129	
				Stomach cancer			161	
			Metal furnacemen	All causes			113	
				Lung cancer			155	
Enterline & McKiever (1963)	US vital statistics 1950		Metal moulders	Lung cancer	34		227	
				Lung cancer	34		170	PMR
Milham (1976a,b)	Washington State death records 1950-1971		Metal moulders	All cancers	72	67	108	PMR
				Stomach	11	7	166	
				Respiratory system	21	16	135	
Petersen & Milham (1980)	California State death records		Metal moulders	Lung cancer	10	5	186	PMR

[a]SMR, unless otherwise specified; SMRs indicated by an asterisk (*) are statistically significant.

[b]PMR, proportionate mortality ratio

[c]From Office of Population Censuses and Surveys (1978)

deaths were calculated both from the age-specific death rates (SMR analysis) and from the proportion of lung cancer deaths, out of all deaths, in the same population (PMR analysis). Since PMRs are based only on death certificates, they were thought useful in order to avoid differences in reports of occupations between census schedules and death records. Among 'metal moulders' (62 905 estimated population), 34 deaths from lung cancer were observed, yielding a SMR of 227 and a PMR of 170, both statistically significant.

Death records for males aged 20 years or more during the period 1950-1971 in Washington State, USA, were analysed by Milham (1976a,b). Some 300 000 death certificates were considered from which occupations were obtained. A standardized proportionate mortality analysis was carried out. Among 'metal moulders', 72 cancers were observed compared to 67 expected; 11 deaths were from stomach cancer compared to seven expected; 21 deaths were from cancer of the respiratory system compared to 16 expected.

Petersen and Milham (1980) carried out a similar analysis for the period 1959-1961 in the State of California and they found among 'metal moulders' 10 lung cancer deaths compared to five expected. Further analysis of the Washington State death records in the same period yielded 13 deaths from lung cancer observed compared to eight expected.

[Surveys of mortality statistics suffer from many limitations. For instance, analysis is cross-sectional, since the number of deaths is related to denominators drawn from the census with no follow-up of exposed subjects. Information sources for numerators and denominators are independent (except when PMRs are calculated); information on causes of death in a given occupation is, in fact, obtained from death certificates, while information on persons in the same occupation is based upon census figures. Occupation recorded on death certificates may correspond to the last occupation. Occupational categories identified have poor specificity and are often much broader than desired.]

(*c*) *Cohort mortality studies* (See Tables 21 and 22)

In 1962, an extensive mortality study on steel workers in seven US steel plants was initiated by the US Public Health Service and the University of Pittsburgh, Graduate School of Public Health (Lloyd & Ciocco, 1969). All men employed in 1953 in any of the seven plants were admitted to the study, and their vital status was ascertained as of 31 December 1961. Information on employment and vital status was obtained for 99.8% of the men in the cohort. Complete work histories from the beginning of employment through 1961 were obtained from plant personnel records for each individual in the study; all causes of death included in the study were confirmed by death certificates. Lloyd *et al.* (1970) examined the relationship between mortality from all causes and specified work areas. Calculation of expected deaths within each work area was based on the mortality experience of the total steelworker population under study. Among the 1143 men employed in 1953 in the foundry area, 90 deaths from all causes were observed whereas 101.1 were expected. The foundry turned out to be one of the 28 work areas exhibiting a mortality lower than expected; in 25 remaining areas, mortality from all causes was greater than expected. In a later report (Lerer *et al.*, 1974), updated results of the study, extended so as to cover the period 1953-1966, were presented. The paper was concerned mainly with the mortality patterns among crane operators; however, a few data regarding the foundry area were given. There were 1962 white men ever employed in the foundry area of the seven plants under study during the period 1953-1966, and, among them, 326 deaths from all causes were observed compared to 326.5 expected. As regards lung cancer, 23 deaths occurred while 20.9 would have been expected. [No analysis by duration of employment, length of follow-up, latency, etc., was carried out among foundry workers. The occupational classification 'foundry workers' appears to be too broad.]

Table 21. Epidemiological studies of iron and steel foundry workers: Cohort studies

Reference	Study subjects	Comparison population	Period of follow-up	Cause of death	Occupation/ exposure	Deaths: Observed	Deaths: Expected	SMR[a]	Comments
Lloyd *et al.* (1970)[b]	1143 foundrymen employed in 1953 at seven steel plants	Total group of steel workers in the same plants (58 128)	1953-1962	All causes	Steel foundry	90	101.1	89	
Lerer *et al.* (1974)[b]	1962 foundry men ever employed in seven steel plants	Total steel-worker population	1953-1966	All causes	Steel foundry	326	326.5	1.00	RR
				Lung cancer	Steel foundry	23	20.9	1.11	
Breslin (1979)[b]	2167 foundrymen in steel plants	Total steelworker population in the same plants	1953-1970	All causes	Cohort A	218	225.5	0.96	RR
					Cohort B	461	461.4	1.00	A, first job in 1953
					Cohort C	271	271.8	1.00	was in the foundry;
				All cancers	Cohort A	62	48.3	1.32*	B, ever employed
					Cohort B	104	98.2	1.07	in the foundry
					Cohort C	69	58.3	1.21	through 1953;
				Lung cancer	Cohort A	19	14.4	1.34	C, employed in the
					Cohort B	28	29.1	0.96	foundry for >5
					Cohort C	21	17.3	1.23	years through 1953
				Genito-urinary cancer	Cohort A	12	5.7	2.24*	
					Cohort B	17	11.5	1.53	
					Cohort C	14	7.4	2.01*	
Redmond *et al.* (1981)[b]		Foundry workers	1953-1975	All cancers	Cohort A			1.27*	RR
					Cohort B			1.12	
					Cohort C			1.27	
				Lung cancer	Cohort A			1.39	
					Cohort B			0.99	
					Cohort C			1.30	
				Genito-urinary cancer	Cohort A			1.99*	
					Cohort B			1.81*	
					Cohort C			2.40*	
				Prostatic cancer	Cohort A			2.16*	
					Cohort B			1.91*	
					Cohort C			2.20*	
				Kidney cancer	Cohort A			-	
					Cohort B			2.55*	
					Cohort C			3.68*	
Koskela *et al.* (1976)	3876 foundry workers with at least 3 months' exposure	Finnish male population, 1967	1950-1973	All causes	Iron, steel and non-ferrous foundry	224	249.4	90	Overall mortality shifted toward younger ages

Table 21 (contd)

Reference	Study subjects	Comparison population	Period of follow-up	Cause of death	Occupation/ exposure	Deaths			Comments
						Observed	Expected	SMR[a]	
				Lung cancer	Iron, steel and non-ferrous foundry	21	13.9	151	Higher SMR for workers exposed >5 years than <5 years
				Lung cancer	Iron, steel and non-ferrous foundry	21	12.1*		PMR
				Lung cancer	Iron foundries	10	3.7	270*	Only workers exposed for at least 5 years
					Steel foundries	0	1.5		
					Non-ferrous foundries	1	0.7		
				Lung cancer	High dust >5 years	8	2.9*		
					Slight dust <5 years	9	5.3		
				Lung cancer	Moulders, coremakers	6	2.6		
					Casters, furnacemen	5	2.1		
					Fettlers	3	2.7		
					Labourers	5	3.6		
Tola *et al.* (1979)	3425 foundrymen with at least 1 year's employment between 1918 and 1972	Finnish male population	1953-1976	Lung cancer	Iron foundry	51	35.3*		PMR
Gibson *et al.* (1977)	1542 foundrymen aged >45 years in 1967	Metropolitan Toronto population	1967-1976	All causes	Steel casting division	95	105,9	90	
					Cranes	9	7.5	121	
					Finishing	30	26.9	115	
					Moulding	15	19.2	78	
					Coremaking	14	12.0	116	
				All cancers	Steel casting division	37	26.7	138	
					Cranes	6	1.8	324	
					Finishing	12	7.0	171	
					Moulding	7	5.0	141	
					Coremaking	4	3.1	130	

				Lung cancer	Steel casting division	21	8.4	250*	Suggestion of a dose-response relationship for lung cancer
					Cranes	4	0.6	714*	
					Finishing	7	2.2	314	
					Moulding	4	1.6	255	
					Coremaking	2	1.0	208	
Decouflé & Wood (1979)	2861 foundrymen employed for at least 1 month	US male population (whites)	1938-1967		Grey iron foundry				
				All causes	>1 month >5 years	247 100	301.3 125.0	0.82* 0.80*	
				All cancers	>1 month >5 years	55 23	49.4 20.4	1.11 1.13	
				Respiratory cancer	>1 month >5 years	17 5	12.6 4.9		
				Digestive cancer	>1 month >5 years	16 9	18.0 7.8		
		US male population (non-whites)		All causes	>1 month >5 years	182 82	315.9 125.4	0.58* 0.65*	
				All cancers	>1 month >5 years	35 18	39.6 17.2	0.88 1.05	
				Respiratory cancer	>1 month >5 years	12 7	10.5 4.5		
				Digestive cancer	>1 month >5 years	14 8	15.2 6.7		
	Subcohort of workers with at least 5 years' employment prior to 1938	US male population		Digestive cancer		14	7.4*		
				Respiratory cancer		8	4.0		
Egan *et al.* (1979)	3013 members of moulders' unions; death benefits were paid between 1971 and 1975	US male population, 1973	1971-1975	All cancers	Foundry workers	501	453.7	110*	PMR; death certificates available for 2734 (92%) subjects
				Lung cancer	Foundry workers	208	141.6	147*	
				Cancer of buccal cavity	Foundry workers	11	7.0	158	
				Bladder cancer	Foundry workers	32	27.6	116	
Egan-Baum *et al.* (1981)	3013 members of moulders' unions, as Egan *et al.* (1979)	US male population, 1973	1971-1975	All cancers	Iron, steel, non-ferrous foundries				PMR; updating of Egan *et al.* (1979) study (death certificates obtained for 99.2%)

Table 21 (contd)

Reference	Study subjects	Comparison population	Period of follow-up	Cause of death	Occupation/ exposure	Deaths			Comments
						Observed	Expected	SMR[a]	
					Whites	545	497.6	110*	
					Blacks	86	69.3	124*	
				Lung cancer	Whites	224	155.2	144*	
					Blacks	39	22.1	176*	
				Cancer of buccal cavity	Whites	11	7.6	144	
					Blacks	0	1.3		
				Bladder cancer	Whites	39	30.3	129	
					Blacks	4	2.4		
Fletcher & Ades (1984)	10 250 foundry workers employed for at least 1 year who started work between 1946 and 1965	England and Wales population	1946-1978 or 1980	All causes	Steel foundry	1850	1756		
				All cancers	Steel foundry	507	446.5		
				Lung cancer	Steel foundry	253	184.7		
				All causes	Foundry *per se*	815	766.2	106	
					Fettling shop	392	354.9	110.5	
					Others	643	634.4	101.4	
				All cancers	Foundry *per se*	244	195.1	125*	
					Fettling shop	109	92.1	118	
					Others	154	159.3	97	
				Lung cancer	Foundry *per se*	115	80.8	142*	
					Fettling shop	66	38.2	173*	
					Others	72	65.7	110	
				Lung cancer	Furnace repairmen	19	9.4	203*	
					Labourers	41	29.6	139	
					Fettlers	32	16.4	195*	
					Heat treatment workers	8	2.3	356*	
					Maintenance fitter's mates	17	7.6	225*	

[a]SMRs indicated by an asterisk (*) are statistically significant.

[b]Continuations of the study initiated by Lloyd and Ciocco (1969)

Table 22. Epidemiological studies of iron and steel foundry workers: Case-control studies

Reference	Cases	Controls	Period of case inception	Relevant results	Comments
Tola *et al.* (1979)[a]	51 lung cancer deaths in a cohort of 3425 iron foundry workers	(1) Workers in the cohort deceased but not from cancer or accidents (3 controls for each case) (2) Same, but control alive at the time the case died (1 control for each case)	1953-1976	(1) and (2) Occupations more common among cases: floor moulders, coremakers, casters and fettlers. Heavy exposure to polynuclear aromatic hydrocarbons more frequent among cases than among controls	(1) and (2) Differences not statistically significant
		(3) 1303 live workers in the cohort		(3) Observed frequencies of casters and floor moulders among cases significantly higher than expected from control occupational distribution	(3) Unusual approach, not classical case-control
Egen-Baum *et al.* (1981)[a]	113 lung cancer deaths among foundrymen of 19 local branches of union	249 deaths from the same branches not due to cancer or nonmalignant respiratory disease	1971-1976	62% of the cases had worked in iron foundries compared to 50% of the controls	Only 65% response rate for working history
				OR[b] for iron *versus* steel and non-ferrous, 2.36 (CL[c], 1.01-5.53)[d] OR for iron *versus* non-ferrous, 2.67 OR for steel *versus* non-ferrous, 1.25	ORs for foundry workers who died before the age of 65 years. No relevant differences for those of older age
Blot *et al.* (1983)	335 lung cancer cases in an industrialized area	332 deaths from causes other than respiratory cancer and disease and suicide	1974-1977	OR for exposure in steel foundry, 7.1 (CL, 1.2-42.3)	Population-based case-control study. ORs exceeded 1.0 in each steel industry job category, the one for foundry being the highest
Neuberger *et al.* (1982)	1248 dust-exposed workers born before 1911	1160 controls	1950-1960	Lung cancer: 175 exposed, 130 controls	

[a]Carried out within the data base of proportionate mortality studies

[b]OR, odds ratio

[c]CL, confidence limits

[d]Statistically significant

This population of steel foundry workers was further studied by Breslin (1979). The follow-up covered the period 1953-1970, and vital status information became available for each man in the original cohort of 58 828. According to the complete set of work histories obtained, 2167 men had worked at some time in the foundry department as of the end of 1953. Three cohorts of white foundry workers were defined (not mutually exclusive): (A) employees whose first job in 1953 was in the foundry; (B) employees who had ever worked in the foundry through 1953; (C) employees with five years or more experience in a foundry job through 1953. Underlying causes of death were obtained from death certificates. Expected deaths were computed on the basis of the mortality rates of the total steelworker population under study. The number of deaths among non-white workers was too small, and results are not presented. In the three cohorts A, B and C, the overall mortality was slightly lower than expected. In cohort A, deaths due to cancer exhibited a statistically significant excess (62 observed; 48.3 expected); also in cohorts B and C, more cancer deaths were observed than expected (104/98.2 and 69/58.3, respectively). In cohort A, 19 deaths from lung cancer were observed compared to 14.4 expected. In cohort B, lung cancer deaths were fewer than expected (28/29.1), while in cohort C, they were slightly in excess (21/17.3). A significant excess of deaths from cancers of the genito-urinary system was observed in cohorts A (12/5.7) and C (14/7.4), and an excess was also present in cohort B (17/11.5). Most of the excess can be attributed to prostatic cancer, but consistent although non-significant excesses were also observed for bladder and for kidney cancers. In cohorts B and C, a slight excess of mortality from lymphatic and haematopoietic neoplasms (11/8.4 and 6/4.7, respectively) was observed. Although 10 occupational groups were identified within the foundries, the numbers of subjects were too small to allow statements about the relation of specific jobs to cancer risks.

When the mortality of this population was followed up to 1975 (Redmond *et al.*, 1981), a significant excess of deaths from all cancers was seen in white workers of cohorts A and C of Breslin (1979) (relative risk, 1.27; $p < 0.05$); and the excess of cancer of the genito-urinary system continued to be significant in all of the three cohorts defined by Breslin (1979), with relative risks of 1.99 ($p < 0.05$), 1.81 ($p < 0.01$) and 2.40 ($p < 0.01$), respectively. The relative risks for cancer of the prostate were 2.16 ($p < 0.05$), 1.91 ($p < 0.05$) and 2.20 ($p < 0.05$), respectively. In addition, the excess of deaths from cancer of the kidney among white workers became significant in cohorts B and C (relative risks, 2.55, $p < 0.05$, and 3.68, $p < 0.01$). The relative risks for lung cancer were similar to those seen previously: A, 1.39; B, 0.99; C, 1.30, none of which was significant.

In Finland, a comprehensive study regarding the work environment and health status of employees in steel, iron and non-ferrous foundries was carried out between 1972-1975. A mortality study, the methods and results of which are given by Koskela *et al.* (1976), was part of this project. Of the 15 401 men employed between 1 January 1950 and 31 December 1972 in 20 selected foundries, all of the 1233 men exposed for at least five years were admitted to the study. In addition, samples of foundry workers with shorter lengths of employment were included in the study: all of the 629 workers with a working history of three to five years; 1042 workers sampled from those exposed for one to three years; and 972 from those exposed for three months to one year. The number of workers in the study thus became 3876. The study covered the period 1950-1973 inclusive. Individual work histories were collected from the employers' records. Workers were grouped into four occupation categories: casters and furnacemen, fettlers, moulders and coremakers, and unskilled labourers. The vital status of each worker was ascertained through the Population Data Register of the Social Insurance Institution, which includes the entire Finnish population. When, on a death certificate, cancer was mentioned as the underlying cause of death, more detailed information was obtained from the Finnish Cancer Registry; 98.7% of the workers in the cohort were traced in this way. The mortality of the cohort was compared with that of the general Finnish male population for the year 1967. This year was selected because most of the deaths occurred at the end of the 1960s. The overall mortality of the cohort of foundry workers was lower than expected (SMR, 90). The ratio remained lower than 100 when people with less than and those with more than five years of exposure were analysed separately: the results did not change when only workers who had entered the cohort between 1950-1960 were considered. The overall mortality failed to exhibit particular trends when analysed by duration of exposure. No 'remarkable' difference regarding mortality from all causes was observed when comparing different types of foundry (iron, steel, non-ferrous). Among the entire cohort, 21 deaths from lung cancer occurred while 13.9 would have been expected (SMR, 151). Workers were also classified according to length of exposure, as obtained from personnel records. For those exposed for less than five years, 10 lung cancer deaths were observed compared to 7.9 expected based on mortality rates of the Finnish population; for those exposed for more than five years, 11 lung cancer deaths were observed compared to 5.9 expected. Among the latter group of workers, when only 'typical' foundry occupations were taken into account (excluding patternmakers, inspectors, maintenance, etc.), the ratio became higher, 10/4.8. Lung cancer deaths that occurred among workers with at least five years of exposure were also analysed according to the type of foundry; 10 of the 11 lung cancer deaths had occurred among workers in iron foundries while only 3.7 were expected (the excess was statistically significant; confidence limits, 130-497). Workers were also classified according to the 'dustiness' of their job. Among those exposed for at least five years in 'high dust' occupations, lung cancer was more frequent (observed deaths, 8; expected, 2.9; statistically significant) than among workers employed in 'slight dust' occupations for less than five years (9/5.3). Finally, an analysis of lung cancer deaths was carried out by occupational group, using a direct standardization method. The following results were found: 'moulders and coremakers', 6/2.6; 'casters and furnacemen', 5/2.1; 'fettlers', 3/2.7; 'labourers', 5/3.6. The detailed occupational histories of the 10 lung cancer cases with more than five years of exposure were compared with those of an equal number of controls matched by age. There were significantly more moulders among cases (5/10) than among controls (0/10). According to the authors, smoking seemed an unlikely explanation for the findings, since, according to data collected from questionnaires, the smoking habits of the current foundry workers did not differ remarkably from those of the Finnish population.

Lung cancer risk among male workers in 13 Finnish iron foundries was investigated more closely in a proportionate mortality study (Tola *et al.*, 1979). All male workers employed for at least one year between 1918-1972 in one of the iron foundries were included in the cohort. These criteria were met by 3425 workers, 1389 of whom had already been included in the previous study. Mortality was examined for the period 1953-1976 and causes of death ascertained through death certificates. Numbers expected were calculated from the age- and period-specific proportions of lung cancer deaths in the Finnish male population; 51 deaths from lung cancer were observed compared to 35.3 expected, a statistically significant excess. A case-control study was set up in order to investigate thoroughly the relationship between lung cancer deaths and occupational history. Controls for the 51 cases were selected from workers in the iron foundries under study according to two different approaches: (1) three dead controls matched for age and (2) one control of the same age as the case, alive at the time of death of the case and who had entered the foundry within the same year. Another approach was attempted in which the distribution of occupations among cases was compared with an 'expected' distribution calculated from a group of 1303 foundry workers (3). Information on occupation and exposures was collected by means of questionnaires sent to next-of-kin and then checked at the workplaces, and polynuclear aromatic hydrocarbons (PAHs) were measured in the work environment. 'Casters, knockoutmen, moulders' were classified as having 'heavy' PAH exposure; 'coremakers' as having 'some' exposure; 'fettlers' as having 'low' exposure; for other occupations the exposure was considered as 'miscellaneous or undefined'. Comparisons between different jobs and levels of exposure revealed no significant difference, except that using approach (3), where the number of lung cancer cases in which employment had been 'floor moulder' or 'caster' was significantly higher than expected. This finding could not be explained by differences in smoking habits or socioeconomic grouping.

In Canada, Gibson *et al.* (1977) studied lung cancer mortality in a cohort of foundry workers in an integrated steel mill. All plant employees and retired workers alive in 1967 and aged 45 years or more were identified. Of the 1542 individuals who met the above criteria, 439 were classified as 'foundry men' who had worked for at least five years in the foundry during their employment prior to 1967; 1103 workers, with at least five years of employment, but less than five years in the foundry, were classified as 'non-foundry'. Mortality was investigated during the period 1967-1976 inclusive. Vital status was ascertained for all but six workers. Among the workers in the original cohort, 95 deaths occurred in the foundry group (37 from cancer) and 200 in the 'non-foundry' group (49 from cancer). The total number of lung cancer deaths was 21 in the foundry group and 11 in the 'non-foundry' group. Lung cancer deaths were significantly more frequent among foundry workers than among 'non-foundry' workers in the age groups 45-54 years and 55-64 years but not among those aged 65 years or more. For foundry workers of all ages, the authors estimated a relative risk of 5.01 for lung cancer in comparison to 'non-foundry' workers. Expected deaths were calculated on the basis of the 1971 metropolitan Toronto mortality rates. There was a significant excess of deaths from lung cancer for all foundry workers (21 observed, 8.4 expected), while among 'non-foundry' workers observed deaths were fewer than expected. Analysis by job category showed a statistically significant excess of deaths from lung cancer for crane operators (4/0.6); a non-significant excess was observed among workers in the finishing, moulding and coremaking departments. Analysis according to length of employment showed a significant excess of deaths from lung cancer among foundry workers with exposure of 20 years or more (11/4.2) and among foundry employees whose initial exposure occurred more than 20 years prior to commencement of the study (17/7.5). A preliminary industrial hygiene survey of current levels of particulates, benzene-soluble fraction of total suspended particulates, and metals was carried out during 1976-1977. Benzo[*k*]fluoranthene, benz[*a*]anthracene and pyrene were each present in measurable quantities (>0.001 μg/m³). Nickel and chromium exposure levels were all less than 25 μg/m³, and exposure levels for other materials were well below their current threshold limit values. It was considered difficult to relate any of the current levels of known carcinogens to the excess of lung cancer found in the foundry, however, levels of such compounds may have been considerably higher in the past. It was considered that exposure to pyrolysis products of the organic binder compounds was more likely to have occurred for crane operators, moulders and coremakers. No difference in smoking habits was observed between foundry and 'non-foundry' workers. [The Working Group noted that the definition of workers with less than five years of work in the foundry as 'non-foundry' workers was inappropriate.]

Decouflé and Wood (1979) examined the long-term mortality experience of workers in a grey-iron foundry in the USA. Of 23 698 'blue-collar' workers employed for at least one year in a large industrial plant between 1938-1967, 2861 subjects, who had worked for at least one month in the foundry department before 1 January 1968, were included in the study. Cause- and age-specific mortality rates for the general US male population (whites and non-whites separately) were used to compute expected numbers of deaths. The study covered the period 1 January 1938 to 31 December 1967. Less than 2% of the white workers could not be traced, and about 4% of non-whites were lost to follow-up. Overall mortality was less than expected among both whites and non-whites. Mortality from all cancers was slightly in excess among whites. Analysis by detailed cancer sites failed to show any significantly increased risk. When analysis was confined to foundry workers with at least five years of employment prior to 1938, there was a statistically significant increase in the ratio of observed to expected numbers of deaths from digestive cancer (14/7.4), and a two-fold non-significant increase of deaths from respiratory cancer (8/4.0). [The rather small number of workers in this group limited the interpretation of the findings. No qualitative or quantitative information concerning exposures in the foundry was available to the authors, and no further subdivision of workers into specific jobs was therefore possible.]

Another study of foundry men in the USA was carried out by Egan *et al.* (1979) on 3013 deaths among male members of the International Molders and Allied Workers Union for whom death benefits were paid between 1971-1975. Every deceased worker had been a union member for at least 11 years (which did not necessarily correspond to the length of employment). At the time the study was published, 8% of death certificates had not yet been located. Expected deaths were calculated on the basis of the proportionate mortality of the US male population (30 years old or more) in 1973. Workers could not be classified according to type of foundry; however, more than 50% of the foundries organized by the Union were ferrous foundries. There was a statistically significant excess of deaths from all cancers (501 observed, 453.7 expected) and of deaths from lung, tracheal and bronchial cancer (208/141.6). These results, according to the authors, should be interpreted with caution, since about 30% of the Union members dropped out of the benefit programme for non-payment of dues. [No information on smoking habits was available.]

The above study was later updated (Egan-Baum *et al.*, 1981) when 99.2% of the death certificates of the 3013 deceased Union members had become available. The study confirmed the results obtained previously: statistically significant excesses were seen for both black and white workers for deaths from all cancers (for whites, 545 observed, 498 expected; for blacks, 86/69) and from lung, tracheal and bronchial cancer (whites, 224/155; blacks, 39/22). A case-referent study was carried out within the data base of the mortality study to identify whether differential lung cancer risk existed for iron foundry men compared with other foundry workers. From local branches of the Union in which at least four lung cancer deaths had occurred, 113 white male cases of lung cancer were selected; 249 controls were selected among all other deaths which had occurred in the same branches from causes other than cancer and non-malignant respiratory diseases. They were matched for date of birth and date of death. Information about employment was obtained by questionnaires sent to the branches and supplemented by information on death certificates. This procedure, however, provided information on foundry type for only 65% of the 362 subjects in the study (60% of the cases and 67% of the controls). The results showed that a higher proportion of cases (62%) than controls (50%) had worked in iron foundries. When comparing lung cancer risk in the iron foundry group with that in the steel and non-ferrous foundry group, an odds ratio of 2.36 (95% confidence limits, 1.01-5.53) was obtained in the age range 42-64 years. The lung cancer risk in the older age group was not significantly elevated. [Inadequate employment information precluded analysis by type of job, employment dates, etc. Smoking histories were not available, but smoking was considered to be an unlikely explanation of the findings.]

Fletcher (1983) and Fletcher and Ades (1984) studied the mortality experience of workers in the nine British steel foundries which, of the Steel Castings Research and Trade Association member foundries, had appropriate records and had agreed to cooperate. From plant personnel records, details were taken of all male workers who had had at least one year of employment in the foundry and who had started work sometime between 1946 and 1965. According to the given criteria, 10 250 workers were admitted to the study. Their mortality experience was examined up to the end of 1978. The tracing was successful for all but 231 (2.2%) workers, and a further 173 had left the country. The analyses by length of employment and length of follow-up made use of mortality up to December 1980, by which time 2105 of the cohort members had died and the proportion untraced had fallen to 1.8%. Observed and expected deaths occurring in 1979 and 1980 included in the analysis by length of employment were restricted to those who had already left. In 14 cases, cause of death could not be ascertained. According to the descriptions obtained by personnel officers and supervisors, 25 occupational categories were identified and grouped into three major work areas, i.e., foundry, fettling shop, pattern/machine/maintenance. Each worker was assigned to the category occupied for the longest time. Expected deaths were calculated on the basis of death rates in England and Wales. In the whole cohort, 1850 deaths had occurred during the study period and 1756 were expected. Excesses were observed for deaths from all cancers (507 observed; 446.5 expected) and from lung cancer (253/184.7). A great inter-foundry variation was observed. Overall mortality by work area did not show remarkable deviations from expectations. In the foundry as a whole, statistically significant excesses of deaths were seen for all cancers (244/195.1), lung cancer (115/80.8) and stomach cancer (standardized mortality ratio (SMR), 188). In the fettling shop, there was a significant excess of lung cancer deaths (66/38.2). When lung cancer mortality was analysed by specific occupational groups, the results were as follows. In the foundry area, statistically significant results were obtained for the furnace repair men (19/9.4). In the fettling area, a significant increase of lung cancer mortality was observed among fettlers and grinders (32/16.4) and heat-treatment workers (8/2.3). In the remaining work area, lung cancer mortality was high among machinists (19/12.0) and significantly higher than expected among maintenance fitters' mates (17/7.6). It is worth noting that in this third group of occupations, the fitters and their mates are the only workers who spent a varying but large amount of their time in the foundry and fettling areas. When mortality was looked at by date of entry, the SMR for lung cancer deaths appears to be independent of the date of starting work in the foundry, in spite of the shorter period of follow-up for later entrants. When lung cancer mortality was analysed according to length of employment, there were suggestions of an increasing risk with length of employment, but the increase was not steady. Smoking, region and social class were taken into account, but none was considered to be a satisfactory explanation of the findings.

Blot *et al.* (1983) examined lung cancer mortality among steel workers in a case-control study conducted in two industrialized counties of eastern Pennsylvania, USA. From vital statistics records, 360 primary lung cancer deaths occurring in the period 1974-1977 among white male residents aged 30-79 years were identified; 360 decedents from causes other than respiratory cancer, chronic respiratory disease or suicide, matched for sex, age, race, county and year of death, were selected randomly as controls. Confirmation of cause of death and interviews with next-of-kin were obtained for 335 cases and 332 controls. Each individual in the study was assigned to a 'usual' industrial category, defined as the one in which he had spent at least 15 years; the job category within the industrial category was the one held for the greatest number of years. Among workers in the steel industry as a whole there was a significant odds ratio of 2.2; when adjusted for smoking it was 1.8. The highest odds ratio for lung cancer was obtained for 'foundry workers and mould makers', based on six cases and one control (odds ratio, 7.1; confidence limits, 1.2-42.3). The odds ratio, however, exceeded 1.0 in each of the job categories in the steel industry. [Although lung cancer odds ratios adjusted for age, education, smoking and employment in other industries were calculated, these figures were available only for workers in the steel industry as a whole and not for workers employed in specific departments.]

Neuberger *et al.* (1982) extracted 1630 medical records of male workers with a history of prolonged occupational dust exposure from 20 000 records collected between 1950-1960 during a preventive medical survey in Vienna, Austria. All were born before 1911, and those with dust exposure were matched by age, sex, domicile and smoking habits with 1630 controls for whom occupational exposure to dust 'could be excluded'. Of 3250 subjects traced to the end of 1980, 1248 exposed workers and 1160 controls had died. For lung cancer the numbers were 175 in exposed and 130 in controls - a statistically significant difference most notable in the group over 60 years of age. Iron-foundry workers formed the largest dust-exposed group; their relative risk for lung cancer was much the same as that for other dust workers.

[Considerable interest has been expressed recently in the possible relationship between exposure to silica dust and cancer of the respiratory tract (Goldsmith *et al.*, 1982). This issue was not addressed directly in any of the studies reviewed by the Working Group.]

5. Summary of Data Reported and Evaluation

5.1 Exposures

The iron and steel foundry industry employs approximately two million workers. This industry is diverse in terms of materials and processes, resulting in occupational exposures to a variety of substances. Substantial exposures to silica and carbon monoxide continue to occur in many foundries. Occupational exposures to airborne polynuclear aromatic compounds have also been found, resulting mainly from the thermal decomposition of carbonaceous ingredients commonly added to foundry sand. In addition, some steel foundry workers (e.g., fettlers) are exposed to airborne chromium and nickel compounds. The introduction of new organic binder materials (beginning in the late 1950s) has resulted in new exposures of foundry workers to chemicals, including phenol, formaldehyde, isocyanates and various amines.

5.2 Experimental data

No data on the carcinogenicity to experimental animals of complex mixtures found in the iron and steel founding industry were available.

Overall assessment of data from short-term tests on iron and steel foundry air samples[a]

	Genetic activity			Cell transformation
	DNA damage	Mutation	Chromosomal effects	
Prokaryotes		+		
Fungi/ Green plants				
Insects				
Mammalian cells (*in vitro*)				
Mammals (*in vivo*)				
Humans (*in vivo*)				
Degree of evidence in short-term tests for genetic activity: *Inadequate*				Cell transformation: No data

[a]The groups into which the table is divided and '+', '−' and '?' are defined on pp. 16-17 of the Preamble; the degrees of evidence are defined on pp. 17-18.

Numerous samples taken from the air in various locations in one steel two iron foundries were mutagenic to *Salmonella typhimurium*. No data on cell transformation were available.

5.3 Human data

Chronic respiratory effects such as silicosis, other pneumoconioses and chronic bronchitis occur among foundry workers. Occupational asthma and dermatitis have also been described following the introduction of new chemical binders.

Data derived from reports of mortality statistics in the USA and the UK indicate excess mortality from lung cancer for 'foundry workers', but the definition of this occupational category included many occupational groups. Cancers of the digestive tract sometimes occurred in higher ratios than in the total population.

Analytical cohort epidemiological studies of foundry workers were carried out in a number of countries. Elevated risks of lung cancer (between about 1.5- and 2.5-fold), some of which were statistically significant, were observed in foundry workers when compared to the general population. In two study populations, comparison with other steel workers did not consistently reveal an elevated rate.

The proportion of lung cancers among all deaths was evaluated in some studies, and found to be about 1.5- to 1.8-fold higher than this proportion in the general population; however, there are difficulties associated with the interpretation of proportionate mortality ratios.

None of the cohort studies explicitly controlled for the potentially confounding smoking habits of foundry workers, although in some studies in which questionnaire data were used, the smoking habits of current workers were not significantly different from those of the general population. Other potential biases in these studies could have arisen from the imprecise classification of jobs.

An association between foundry work and lung cancer was observed in one case-control study.

In two studies, in which site-specific cancer deaths among iron and steel-foundry workers were compared with the corresponding rates for the general population, significantly increased risks for cancer of the digestive system were observed: in one, the elevated risk was in the 'digestive system', in the other, it was in 'stomach cancer'.

Results from studies of a single cohort of steel-foundry workers in the USA showed a significantly elevated risk of 'cancer of the genito-urinary system' when compared with the entire steel-worker population under study, the risk being significantly elevated also for some specific sites (prostate and kidney).

5.4 Evaluation[1]

The available epidemiological studies provide *limited evidence* that certain exposures in iron and steel founding are carcinogenic to humans, giving rise to lung cancer. There is *inadequate evidence* that such exposures result in cancers of the digestive system and genito-urinary system.

---

[1]For definitions of the italicized terms, see the Preamble, pp. 15 and 19.

A number of individual compounds for which there is *sufficient evidence* of carcinogenicity have been measured at high levels in air samples taken from certain areas in iron and steel foundries[2].

Taken together, the available evidence indicates that occupational exposures occur in iron and steel founding which are probably carcinogenic to humans.

6. References

Adelstein, A.M. (1972) Occupational mortality: Cancer. *Ann. occup. Hyg.*, *15*, 53-57

Baldwin, V.H. & Westbrook, C.W. (1983) *Environmental Assessment of Melting, Pouring, and Inoculation in Iron Foundries* (*Contract No. 68-02-3152, Task 5; 68-02-3170, Task 58*), Research Triangle Park, NC, US Environmental Protection Agency

Bates, C.E. & Scheel, L.D. (1974) Processing emissions and occupational health in the ferrous foundry industry. *Am. ind. Hyg. Assoc. J.*, *35*, 452-462

Beeley, P.R. (1972) *Foundry Technology*, London, Butterworths

Blot, W.J., Brown, L.M., Pottern, L.M., Stone, B.J. & Fraumeni, J.F., Jr (1983) Lung cancer among long-term steel workers. *Am. J. Epidemiol.*, *117*, 706-716

Bolt, Beranek & Newman Inc. (1978) *A Consultation Service in Industrial Hygiene and Safety in the Foundry Industry* (*Report No. 3744*), Washington DC, US Occupational Safety & Health Administration

Boothe, G.F., Steward-Smith, D., Wagstaff, D. & Dibblee, M. (1980) The radiological aspects of zircon sand use. *Health Phys.*, *38*, 393-398

Breslin, P. (1979) *Mortality among foundrymen in steel mills*. In: Lemen, R. & Dement, J.M., eds, *Dusts and Disease*, Park Forest South, IL, Pathotox Publishers Inc., pp. 439-447

Bryant, D.W., & McCalla, D.R. (1982) *Mutagenicity and lung cancer in a steel foundry environment*. In: Heddle, J.A., ed., *Mutagenicity: New Horizons in Genetic Toxicology*, New York, Academic Press, pp. 89-115

Capodaglio, E., Pozzoli, L., Massola, A., Catenacci, G. & Ghittori, S. (1976) Environmental dust in foundries: Risk of silicosis based on dust measurements (Ital.). *Med. Lav.*, *67*, 454-464

Cockcroft, D.W., Cartier, A., Jones, G., Tarlo, S.M., Dolovich, J. & Hargreave, F.E. (1980) Asthma caused by occupational exposure to a furan-based binder system. *J. Allergy clin. Immunol.*, *66*, 458-463

---

[2]See the Appendix, Table 2, and IARC (1982).

Comité des Associations Européennes de Fonderie (1983) *The European Foundry Industry in 1982* (Ger.), *December 1982, Statement of Commission No. 7*, Paris, irreg. pp. from 63-93

Dahlquist, I. (1981) Contact allergy to colophony and formaldehyde from sand cores. *Contact Dermatitis*, *7*, 167-168

Decouflé, P. & Wood, D.J. (1979) Mortality patterns among workers in a gray iron foundry. *Am. J. Epidemiol.*, *109*, 667-675

Drasche, H. (1976) Specific hazardous compounds in the breathing zone of Croning foundries (Ger.) *Mod. Unfallverhüt.*, *20*, 23-27

Eftax, D.S.P., McKillip, W.J. & Brechter, M. (1977) *Environmental considerations in the usage of no-bake foundry binders*. In: *The Working Environment in Ironfoundries*, Birmingham, British Cotton Industry Research Association, pp. 21-1-21-7

Egan, B., Waxweiler, R.J., Blade, L., Wolfe, J. & Wagoner, J.K. (1979) A preliminary report of mortality patterns among foundry workers. *J. environ. Pathol. Toxicol.*, *2*, 259-272

Egan-Baum, E., Miller, B.A. & Waxweiler, R.J. (1981) Lung cancer and other mortality patterns among foundrymen. *Scand. J. Work Environ. Health*, *7*, Suppl. 4, 147-155

Emory, M.B., Goddman, P.A., James, R.H. & Scott, W.D. (1978) Nitrogen-containing compounds in foundry mold emissions. *Am. ind. Hyg. Assoc. J.*, *39*, 527-533

Enterline, P.E. & McKiever, M.F. (1963) Differential mortality from lung cancer by occupation. *J. occup. Med.*, *5*, 283-290

Falck, K. (1982) *Application of the Bacterial Urinary Mutagenicity Assay in Detection of Exposure to Genotoxic Chemicals*, Academic Dissertation, University of Helsinki, pp. 20-25

Fletcher, A.C. (1983) Lung cancer mortality in a population of English foundry workers (Abstract). *Scand. J. Work Environ. Health*, *9*, 62-63

Fletcher, A.C. & Ades, A. (1984) Lung cancer mortality in a cohort of English foundry workers. *Scand. J. Work Environ. Health*, *10*, 7-16

Gardziella, A. (1982) Production of cores according to SO_2 process. Basis and experiences (Ger.). *Giesserei*, *69*, 81-88

Gerhardsson, G. (1976) Dust prevention in Swedish foundries. *Staub-Reinhalt. Luft*, *36*, 433-439

Gerhardsson, G., Engman, L., Andersson, A., Isaksson, G., Magnusson, E. & Sundquist, S. (1974) *Silikosprojektets Slutrapport. Del 2, Målsättning, Omfattning och Resultat* (Final Report of the Silicosis Project. Sect. 2, Aim, Scope and Result) (*Report of Investigation, AMT 103/74-2*), Stockholm, National Board of Occupational Safety & Health, pp. 165-183

Gibson, E.S., Martin, R.H. & Lockington, J.N. (1977) Lung cancer mortality in a steel foundry. *J. occup. Med.*, *19*, 807-812

Goldsmith, D.F., Guidotti, T.L. & Johnston, D.R. (1982) Does occupational exposure to silica cause lung cancer? *Am. J. ind. Med.*, *3*, 423-440

Goss, H.R. (1980) Blowing of chemically bonded sand molds. *Am. Foundrymen's Soc. Trans.*, *88*, 51-56

Gregory, J. (1970) A survey of pneumoconiosis at a Sheffield steel foundry. *Arch. environ. Health*, *20*, 385-399

Gullickson, R. & Doninger, J.E. (1980) Industrial hygiene aspects of olivine sand use in foundries. *Am. Foundrymen's Soc. Trans.*, *88*, 623-630

Gwin, C.H., Scott, W.D. & James, R.H. (1976) A preliminary investigation of the organic chemical emissions from green sand pyrolysis. *Am. ind. Hyg. Assoc. J.*, *37*, 685-689

Gylseth, B., Gundersen, N. & Langård, S. (1977) Evaluation of chromium exposure based on a simplified method for urinary chromium determination. *Scand. J. Work environ. Health*, *3*, 28-31

Heine, H.J. (1977) Molding with expendable polystyrene patterns. *Foundry Manage. Technol.*, *105*, 84-91

Hernberg, J. (1976) The Finnish foundry project. Background and general methodology. *Scand. J. Work Environ. Health*, *2*, (Suppl. 1), 8-12

Hernberg, S., Kärävä, R., Koskela, R.-S. & Luoma, K. (1976) Angina pectoris, ECG findings and blood pressure of foundry workers in relation to carbon monoxide exposure. *Scand. J. Work Environ. Health*, *2*, (Suppl. 1), 54-63

Hespers, W. (1972) A system of sand with coal free from dust using synthetic products as anthracite moulder (Ger.). *Giesserei*, *59*, 765-770

IARC (1982) *IARC Monographs on the Evaluation of the Carcinogenic Risk of Chemicals to Humans*, Suppl. 4, *Chemicals, Industrial Processes and Industries Associated with Cancer in Humans*, Lyon, pp. 91-93, 167-170

IARC (1983) *IARC Monographs on the Evaluation of the Carcinogenic Risk of Chemicals to Humans*, Vol. 32, *Polynuclear Aromatic Compounds, Part 1, Chemical, Environmental and Experimental Data*, Lyon

Kaiser, C., Kerr, A., McCalla, D.R., Lockington, J.N. & Gibson, E.S. (1980) *Mutagenic material in air particles in a steel foundry*. In: Bjørseth, A. & Dennis, A.J., eds, *Polynuclear Aromatic Hydrocarbons: Chemistry and Biological Effects, 4th International Symposium*, Columbus, OH, Battelle Press, pp. 579-588

Kaiser, C., Kerr, A., McCalla, D.R., Lockington, J.N. & Gibson, E.S. (1981) *Use of bacterial mutagenicity assays to probe steel foundry lung cancer hazard*. In: Cooke, M. & Dennis, A.J., eds, *Polynuclear Aromatic Hydrocarbons: Chemistry and Biological Effects, 5th International Symposium*, Columbus, OH, Battelle Press, pp. 583-592

Kalliomäki, P.-L., Korhonen, O., Mattsson, T., Sortti, V., Vaaranen, V., Kalliomäki, K. & Koponen, M. (1979) Lung contamination among foundry workers. *Int. Arch. occup. environ. Health*, *43*, 85-91

Käräva, R., Hernberg, S., Koskela, R.-S. & Luoma, K. (1976) Prevalence of pneumoconiosis and chronic bronchitis in foundry workers. *Scand. J. Work Environ. Health*, *2* (Suppl. 1), 64-72

Koponen, M., Siltanen, E., Kokko, A., Engström, B. & Reponen, J. (1976) Effect of foundry size on the dust concentration of different work phases. *Scand. J. Work Environ. Health*, *2*, (Suppl. 1), 32-36

Koskela, R.-S., Hernberg, S., Käräva, R., Järvinen, E. & Nurminen, M. (1976) A mortality study of foundry workers. *Scand. J. Work Environ. Health*, *2*, (Suppl. 1), 73-89

Lerer, T.J., Redmond, C.K., Breslin, P.P., Salvin, L. & Rush, H.W. (1974) Long-term mortality study of steel workers. VII. Mortality patterns among crane operators. *J. occup. Med.*, *16*, 608-614

Lloyd, J.W. & Ciocco, A. (1969) Long-term mortality study of steelworkers. I. Methodology. *J. occup. Med.*, *11*, 299-310

Lloyd, J.W., Lundin, F.E., Jr, Redmond, C.K. & Geiser, P.B. (1970) Long-term mortality study of steelworkers. IV. Mortality by work area. *J. occup. Med.*, *12*, 151-157

Logan, W.P.D. (1982) *Cancer Mortality by Occupation and Social Class 1851-1971* (*IARC Scientific Publications No. 36/Studies on Medical and Population Subjects No. 44*), London, Her Majesty's Stationary Office/Lyon, International Agency for Research on Cancer, pp. 201-211

Lukacek, G. & Heine, H.J. (1982) Using core and mold coatings - A new understanding. *Foundry Manage. Technol.*, *110*, 22-26

MacBain, G., Cole, C.W.D. & Shepherd, R.D. (1962) Pneumoconiosis in a group of large iron and light alloy foundries. *Trans. Assoc. ind. med. Off.*, *12*, 17-28

McLaughlin, A.I.G. (1957) Pneumoconiosis in foundry workers. *Br. J. Tuberc. Dis. Chest*, *51*, 297-309

McLaughlin, A.I.G. & Harding, H.E. (1956) Pneumoconiosis and other causes of death in iron and steel foundry workers. *Arch. ind. Health*, *14*, 350-378

Meyer, P.B. (1973) Dust measure in Dutch iron foundries (Ger.). *Staub-Reinhalt. Luft*, *33*, 76-79

Milham, S., Jr (1976a) Cancer mortality patterns associated with exposure to metals. *Ann. N.Y. Acad. Sci.*, *271*, 243-249

Milham, S., Jr (1976b) *Occupational Mortality in Washington State, 1950-1971* (*HEW Publ. No. (NIOSH) 76-175-A*), Cincinnati, OH, National Institute for Occupational Safety and Health

Miske, J.C. (1978) The foundry industry - A look ahead. *Foundry Manage. Technol.*, *100*, 38-48

Morgan, W.K.C. & Seaton, A. (1975) *Occupational Lung Disease*, Philadelphia, W.B. Saunders, pp. 235-239

Morrison, S.L. (1957) Occupational mortality in Scotland. *Br. J. ind. Med.*, *14*, 130-132

Mosher, G.E. (1980) Nickel and chromium exposures in foundries melting/pouring alloys containing low or trace levels of nickel or chrome. *Am. Foundrymen's Soc. Trans.*, *88*, 515-518

National Finnish Board of Occupational Safety and Health (1981) *Airborne Contaminants in the Work Places* (*Safety Bull. No. 3*), Helskinki, p. 18

National Institute for Occupational Safety and Health (1978) *An Evaluation of Occupational Health Hazards Control Technology for the Foundry Industry* (*DHEW (NIOSH) Publication No. 79-114*), Cincinnati, OH, pp. 3, 6, 7

Neuberger, M., Kundi, M., Haider, M. & Gründorfer, W. (1982) *Cancer mortality of dust workers and controls - Results of a prospective study*. In: *Prevention of Occupational Cancer-International Symposium* (*Occupational Safety & Health Series No. 46*), Geneva, International Labour Office, pp. 235-241

Novelli, G. & Rinaldi, A. (1977) *Thermic decomposition products of green sand additives*. In: *44th International Foundry Congress, Florence 11-14 September 1977*, Zurich, Comité International des Associations Techniques de Fonderie, pp. 186-196

Novelli, G. & Rinaldi, A. (1979) Further contributions to the study of coal dust for foundries. *Giesserei*, *66*, 480-482

Office of Population Censuses and Surveys (1978) *Occupational Mortality. The Registrar General's Decennial Supplement for England and Wales, 1970-72*, London, Her Majesty's Stationery Office, pp. 112, 121, 135, 149

Oudiz, J., Brown, J.W., Ayer, H.E. & Samuels, S. (1983) A report on silica exposure levels in the United States foundries. *Am. ind. Hyg. Assoc. J.*, *44*, 374-376

Palmer, W.G. & Scott, W.D. (1981) Lung cancer in ferrous foundry workers: A review. *Am. ind. Hyg. Assoc. J.*, *42*, 329-341

Palmer, W.G., Moorman, W., Stettler, L., James, R. & Scholz, R. (1981) Analyses of effluents collected from four types of iron casting molds for use in carcinogenesis bioassays. *Am. Foundrymen's Soc. Trans.*, *89*, 653-658

Petersen, G.R. & Milham, S., Jr (1980) *Occupational Mortality in the State of California 1959-61* (*DHEW (NIOSH) Publ. No. 80-104*), Cincinnati, OH, National Institute for Occupational Safety and Health

Pompeii, C.W. (1983) *Lead problems in gray iron foundries*. In: *American Industrial Hygiene Conference, Philadelphia, May 22-27*, Abstract No. 453, p. 180

Radia, J.T. (1977) *Environmental considerations in the use of coal-dust substitutes*. In: *The Working Environment in Ironfoundries*, Birmingham, British Cotton Industry Research Association, pp. 18-1-18-12

Rassenfoss, J.A. (1977) Mold materials for ferrous castings. *Mod. Cast.*, *67*, 45-48

Redmond, C.K., Wieand, H.S., Rockette, H.E., Sass, R. & Weinberg, G. (1981) *Long-term Mortality Experience of Steelworkers* (*DHHS (NIOSH) Publ. No. 81-120*), Cincinnati, OH, National Institute for Occupational Safety and Health

Rosenberg, C. & Pfäffli, P. (1982) A comparison of methods for the determination of diphenylmethane diisocyanate (MDI) in air samples. *Am. ind. Hyg. Assoc. J.*, *43*, 160-163

Rüttner, J.R. (1954) Foundry workers' pneumoconiosis in Switzerland (Anthracosilicosis). *Arch. ind. Hyg.*, *9*, 297-305

Schimberg, R.W. (1981) *Industrial hygienic measurements of polycyclic aromatic hydrocarbons in foundries*. In: Cooke, M. & Dennis, A.J., eds, *Polynuclear Aromatic Hydrocarbons - Chemical Analysis and Biological Fate, 5th International Symposium*, Columbus, OH, Battelle Press, pp. 755-762

Schimberg, R.W., Pfäffli, P. & Tossavainen, A. (1978) Profile analysis of polycyclic aromatic hydrocarbons in iron foundries (Ger.). *Staub-Reinhalt. Luft*, *38*, 273-276

Schimberg, R.W., Pfäffli, P. & Tossavainen, A. (1980) Polycyclic aromatic hydrocarbons in foundries. *J. Toxicol. environ. Health*, *6*, 1187-1194

Schimberg, R.W., Toivonen, E. & Tossavainen, A. (1981a) Polycyclic aromatic hydrocarbons and other hazardous agents in foundry moulding sand with various hydrocarbon carriers (Ger.). *Staub-Reinhalt. Luft*, *41*, 221-224

Schimberg, R.W., Skyttä, E. & Falck, K. (1981b) Exposure of foundry workers to mutagenic polycyclic aromatic hydrocarbons (Ger.) *Staub-Reinhalt. Luft*, *41*, 421-424

Schneider, P. (1976) The foundry industry in Brazil (Ger.). *Giesserei*, *63*, 73-78

Scott, W.D., James, R.H. & Bates, C.E. (1976) Foundry air contaminants from green sand molds. *Am. ind. Hyg. Assoc. J.*, *37*, 335-344

Scott, W.D., Bates, C.E. & James, R.H. (1977) Chemical emissions from foundry molds. *Am. Foundrymen's Soc. Trans.*, *85*, 203-208

Shestopal, V.M. (1976) The foundry industry in the USSR. *Foundry Manage. Technol.*, *104*, 46-54

Siltanen, E., Koponen, M., Kokko, A., Engström, B. & Reponen, J. (1976) Dust exposure in Finnish foundries. *Scand. J. Work Environ. Health*, *2* (Suppl. 1), 19-31

Skyttä, E., Schimberg, R. & Vainio, H. (1980) Mutagenic activity in foundry air. *Arch. Toxicol.*, *Suppl. 4*, 68-72

Southern Research Institute (1979) Binder decomposition during pouring and solidification of foundry castings, Part II: Particulate emissions from foundry molds. *Am. Foundrymen's Soc. int. Cast Metals J.*, June, 14-15

Standke, W. & Buchen, W. (1982) Foundry in Japan - Impressions of study journeys (Ger.). *Giesserei*, *69*, 269-276

Stettler, L.E., Gorski, C.H., Platek, F., Stoll, M. & Niemeier, R.W. (1981) Physical and chemical analyses of foundry sands. *Am. Foundrymen's Soc. Trans.*, *89*, 141-156

Swanston, C. (1950) The iron and steel industry. *Lancet*, *i*, 191-197

Toeniskoetter, R.H. (1981) *Urethane foundry binders - An industrial hygiene appraisal.* In: *Proceedings of the Symposium on Occupational Health Hazard Control Technology in the Foundry and Secondary Non-Ferrous Smelting Industries* (*NIOSH Publication No. 81-114*), Cincinnatti, OH, National Institute for Occupational Safety and Health, pp. 142-152

Toeniskoetter, R.H. & Schafer, R.J. (1977) *Industrial hygiene aspects of the use of sand binders and additives.* In: *The Working Environment in Ironfoundries*, Birmingham, British Cotton Industry Research Association, pp. 19-1-19-28

Tola, S. (1980) Epidemiology of lung cancer in foundries. *J. Toxicol. environ. Health*, *6*, 275-280

Tola, S., Kilpiö, J., Virtamo, M. & Haapa, K. (1977) Urinary chromium as an indicator of the exposure of welders to chromium. *Scand. J. Work Environ. Health*, *3*, 192-202

Tola, S., Koskela, R.-S., Hernberg, S. & Järvinen, E. (1979) Lung cancer mortality among iron foundry workers. *J. occup. Med.*, *21*, 753-760

Tossavainen, A. (1976) Metal fumes in foundries. *Scand. J. Work Environ. Health, 2* (Suppl. 1), 42-49

Tossavainen, A. & Schimberg, R. (1981) *Comparison of sand additives as a source of benzo[a]pyrene emissions in foundries.* In: *Proceedings of the 2nd World Congress of Chemical Engineering, Montreal, Canada, Oct. 4-9*, Vol. 6, pp. 132-135

Tossavainen, A., Nurminen, M., Mutanen, P. & Tola, S. (1980) Application of mathematical modelling for assessing the biological half-times of chromium and nickel in field studies. *Br. J. ind. Med.*, *37*, 285-291

Tubich, G.E. (1964) New materials and processes create new liabilities for the foundry. *Ind. Med. Surg.*, *33*, 79-85

Tubich, G.E. (1981) *The foundry. Its real potential health hazards.* In: *Proceedings of the Symposium on Occupational Health Hazard Control Technology in the Foundry and Secondary Non-Ferrous Smelting Industries* (*NIOSH Publication No. 81-114*), Cincinnati, OH, National Institute for Occupational Safety and Health, pp. 3-13

Tubich, G.E., Davis, I.H. & Bloomfield, B.D. (1960) Occupational health studies of the shell-molding process. *Arch. ind. Health*, *21*, 424-444

Turner, H.M. & Grace, H.G. (1938) An investigation into cancer mortality among males in certain Sheffield trades. *J. Hyg.*, *38*, 90-103

Ulfvarson, U. (1981) Survey of air contaminants from welding. *Scand. J. Work Environ. Health*, *7* (Suppl. 2), 1-28

Vanhoorne, M., Dams, R., Bressers, J. & van Peteghem, C. (1972) Smoke of the trigger process in the production of nodular iron and its possible effects on man. *Int. Arch. Arbeitsmed.*, *29*, 102-118

Verma, D.K., Muir, D.C.F., Cunliffe, S., Julian, J.A., Vogt, J.H., Rosenfeld, J. & Chovil, A. (1982) Polycyclic aromatic hydrocarbons in Ontario foundry environments. *Ann. occup. Hyg.*, *25*, 17-25

Virtamo, M. & Tossavainen, A. (1976a) Carbon monoxide in foundry air. *Scand. J. Work Environ. Health*, *2* (Suppl. 1), 37-41

Virtamo, M. & Tossavainen, A. (1976b) Gases formed from furan binding agents. *Scand. J. Work Environ. Health*, *2* (Suppl. 1), 50-53

Virtamo, M., Tossavainen, A. & Ruishalme, J. (1975) *Valimoaineet* (Finn.) (Compounds in the Foundries) (*Research Series 109*), Helsinki, Institute of Occupational Health, pp. 79, 91

Woodliff, E.E. (1971) 75 years progress...in foundry sands and additives. *Mod. Cast.*, *61*, 57-62

Zdražil, J. & Pícha, F. (1963) Cancerogenic substances - 3,4-benzpyrene - in moulding sand mixtures and foundry dust (Czech.). *Prac. Lék.*, *15*, 207-211

Zdražil, J. & Pícha, P. (1965) Carcinogenic hydrocarbons - 3,4-benzpyrene in foundries (Czech.). *Slevarenství 13*, 198-199

Zimmerman, R.E. & Barry, J.M. (1976) Determining crystalline silica compliance using respirable mass. *Am. Foundrymen's Soc. Trans.*, *84*, 15-20

APPENDIX

APPENDIX

Table 1. Chemicals used or produced in the industries considered in this volume which have been previously evaluated in the *IARC Monographs*

Chemical	Industry	Evaluation of carcinogenicity
Acrolein	Foundry	No data on humans; inadequate data on animals (Vol. 19, pp. 479-494, 1979)
Amines, aliphatic and aromatic	Coal gasification Coke production Foundry	Aromatic amines evaluated in volumes 4 (1974), 16 (1978) and 27 (1982)
Arsenic, oxides and salts	Coal gasification Coke production	Carcinogenic in humans (Vol. 23, pp. 39-141, 1980; Suppl. 1, pp. 22-23, 1979; Suppl. 4, pp. 50-52, 1982)
Asbestos	Coal gasification Coke production Foundry Aluminium	Carcinogenic in humans (Vol. 14, 1977; Suppl. 1, p. 23, 1979; Suppl. 4, pp. 52-53, 1982)
Cadmium and cadmium oxides	Coal gasification Foundry	Cadmium and certain cadmium compounds are probably carcinogenic in humans; *sufficient evidence* in animals (Vol. 2, pp. 39-74, 1973; Suppl. 1, pp. 27-28, 1979; Suppl. 4, pp. 71-73, 1982)
Chlorinated hydrocarbons (1,1,1-trichloroethane)	Foundry	No data on humans; no evaluation in animals (Vol. 20, pp. 515-531, 1979)
Chromium and chromium oxides	Foundry	Chromium and certain chromium compounds: carcinogenic in humans; *sufficient evidence* in animals (Vol. 23, pp. 205-233, 1980; Suppl. 1, pp. 29-30, 1979; Suppl. 4, pp. 91-93, 1982)
Ferrochromium, ferromanganese, ferromolybdenum, ferrosilicon, ferrovanadium	Foundry	Ferrochromium (see chromium)
Fluorides	Foundry Aluminium	Inorganic fluorides: no evidence of carcinogenicity in humans; inadequate data on animals (Vol. 27, pp. 237-303, 1982)
Formaldehyde	Foundry Coke production	Inadequate data in humans; *sufficient evidence* in animals (Vol. 29, pp. 345-389, 1982; Suppl. 4, pp. 131-132, 1982)
Hydrocarbons aliphatic, cyclic and aromatic (benzene)	Coke production Foundry	Carcinogenic in humans (Vol. 29, pp. 93-148, 391-398, 1982; Suppl. 1, p. 24, 1979; Suppl. 4, pp. 56-57, 1982)
Isocyanates (e.g., 4,4′-methylenediphenyl diisocyanate)	Foundry	No data on humans or animals (Vol. 19, pp. 314-320, 1979)
Lead and lead oxides	Coal gasification Foundry	Inadequate data on humans and animals (Vol. 23, pp. 325-415, 1980; Suppl. 4, pp. 149-150, 1982)
Oil mists (mineral oils)	Aluminium	*Sufficient evidence* in humans (Vol. 33, pp. 87-168, 1984)
Nickel, metal and oxides	Coal gasification Foundry	Probably carcinogenic in humans; *sufficient evidence* in animals (Vol. 2, pp. 126-149, 1973; Vol. 11, pp. 75-112, 1976; Suppl. 1, p. 38, 1979; Suppl. 4, pp. 167-170, 1982)
Nickel carbonyl	Coal gasification	(See nickel)
Nitrosamines	Foundry	Nitrosamines evaluated in Vol. 17 (1978)
Polynuclear aromatic hydrocarbons	Coal gasification Coke production Foundry Aluminium	No data on humans; evaluations of carcinogenicity in animals are given in General Remarks, Vol. 32 (1983)

Table 2. Chemicals evaluated by an IARC working group in February 1983 (Vol. 32 of the *IARC Monographs*)

Chemicals	Evidence for carcinogenicity in animals	Evidence for activity in short-term tests
Anthanthrene	Limited	Inadequate
Anthracene	No evidence	No evidence
Benz[*a*]acridine	Inadequate	Inadequate
Benz[*c*]acridine	Limited	Inadequate
Benz[*a*]anthracene	Sufficient	Sufficient
Benzo[*b*]fluoranthene	Sufficient	Inadequate
Benzo[*j*]fluoranthene	Sufficient	Inadequate
Benzo[*k*]fluoranthene	Sufficient	Inadequate
Benzo[*ghi*]fluoranthene	Inadequate	Inadequate
Benzo[*a*]fluorene	Inadequate	Inadequate
Benzo[*b*]fluorene	Inadequate	Inadequate
Benzo[*c*]fluorene	Inadequate	Inadequate
Benzo[*ghi*]perylene	Inadequate	Inadequate
Benzo[*c*]phenanthrene	Inadequate	Inadequate
Benzo[*a*]pyrene	Sufficient	Sufficient
Benzo[*e*]pyrene	Inadequate	Limited
Carbazole	Limited	Inadequate
Chrysene	Limited	Limited
Coronene	Inadequate	Inadequate
Cyclopenta[*cd*]pyrene	Limited	Sufficient
Dibenz[*a*,*h*]acridine	Sufficient	Inadequate
Dibenz[*a*,*j*]acridine	Sufficient	Inadequate
Dibenz[*a*,*c*]anthracene	Limited	Sufficient
Dibenz[*a*,*h*]anthracene	Sufficient	Sufficient
Dibenz[*a*,*j*]anthracene	Limited	Inadequate
Dibenzo[*c*,*g*]carbazole	Sufficient	Inadequate
Dibenzo[*a*,*e*]fluoranthene	Limited	No data
Dibenzo[*a*,*e*]pyrene	Sufficient	Inadequate
Dibenzo[*a*,*h*]pyrene	Sufficient	Inadequate
Dibenzo[*a*,*i*]pyrene	Sufficient	Inadequate
Dibenzo[*a*,*l*]pyrene	Sufficient	No data
1,4-Dimethylphenanthrene	Inadequate	Limited
Fluoranthene	No evidence	Limited
Fluorene	Inadequate	Inadequate
Indeno[1,2,3-*cd*]pyrene	Sufficient	Inadequate
1-Methylchrysene	Inadequate	Inadequate
2-, 3-, 4- and 6-Methylchrysenes	Limited	Inadequate
5-Methylchrysene	Sufficient	Limited
2-Methylfluoranthene	Limited	Inadequate
3-Methylfluoranthene	Inadequate	Inadequate
1-Methylphenanthrene	Inadequate	Sufficient
Perylene	Inadequate	Inadequate
Phenanthrene	Inadequate	Limited
Pyrene	No evidence	Limited
Triphenylene	Inadequate	Inadequate

GLOSSARY

As the exact definition of the following terms (in particular job titles) may vary with plant, company, locality and country, this glossary is provided for general guidance, and it is encumbent upon the reader to verify the precise usage of the terms for specific locations under consideration.

Anode-paste plant - the area in a primary aluminium production plant where the paste for Söderberg anodes is prepared

Anode rodding - the process in a primary aluminium production plant using prebake cells, in which rods are removed from spent anodes, reconditioned and mated with new anodes

Beehive oven - a type of coking oven that was used in the earlier part of this century. Its shape was in the form of a beehive, and the volatiles were not collected

Benchman - a worker who cleans the coke-oven door and jambs; he also removes the spilled hot coke from the bench floor

Buss bar - the conducting bar through which electrical current is fed to the anode in primary aluminium production

Butt - the spent anode from a prebake primary aluminium production process

Carbon plant - the area in (adjacent to or remote from) a primary aluminium production plant where prebake anodes are fabricated

Carbon setter - a member of a crew replacing anodes on prebake cells for primary aluminium production

Coke-oven main - a collecting pipe for volatiles located on the topside of a coke-oven battery

Core-box - a moulding box in which a core is made for a casting

Cores - a formed bonded sand insert used inside sand moulds for the production of internal cavities in castings

Crust breaker - the equipment (pneumatic hammer) used to break the crust that forms over the molten electrolyte in the cells used in primary aluminium production

Cyclone - a separator in which particles or droplets are separated from gases or water by centrifugal force

Dross - metallic oxides that rise to the surface of the molten electrolyte in primary aluminium production

Fettling (also referred to as dressing) - a term for the cleaning and tidying of iron and steel castings after shake-out and shotblast. It may involve removal of excess metal (runners, risers, sprues and gates; see diagram) and metal flash as well as rectifying defects such as metal penetration and fins by grinding chipping, etc.

Flex raiser (Rod raiser) - a person removing the bolts holding the electrical connectors to the studs and raising the connectors to the next horizontal row of studs in horizontal-stud Söderberg cells for primary aluminium production

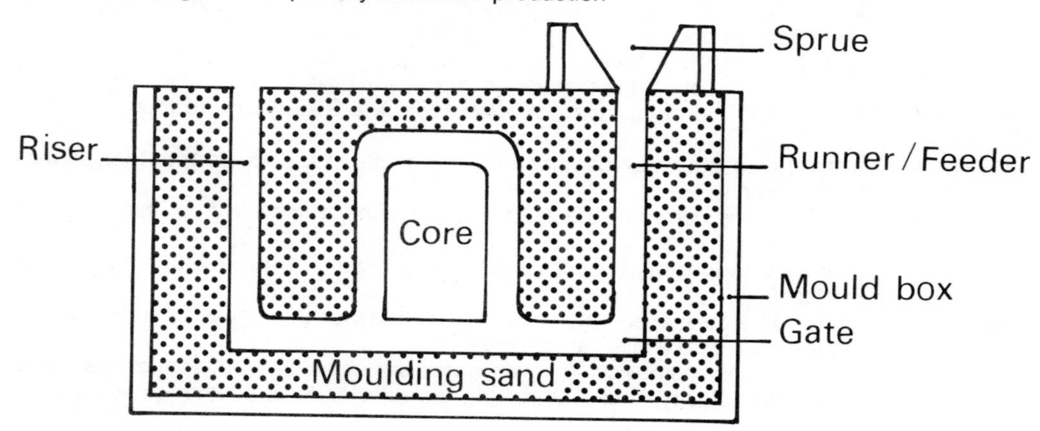

Green carbon - a mixture of coke (usually calcined petroleum pitch) blended with a binder of pitch to form a paste which is pressure moulded to form green anode blocks. These blocks are subsequently baked to form prebake carbon anodes for primary aluminium production

Larry car - a mechanical device that runs on tracks drawing hoppers for the delivery of coal to the charging holes in the top of a coke oven

Luterman - a worker responsible for the sealing of coke-oven doors with a refractory material; in some cases also responsible for sealing of lids and other sources of emissions on the top side

Potman (pot tender) - a person responsible for adding materials to electrolytic cells (pots) used in primary aluminium production and for ensuring their continued operation

Potroom - a structure housing a series of electrolytic cells (pots) for primary aluminium production

Pot - an electrolytic cell used in primary aluminium production for the reduction of alumina to aluminium

Prebake anode - the carbon anode (prepared from green carbon) used in prebake cells for primary aluminium production

Prebake cell (prebake pot) - an electrolytic cell used in primary aluminium production in which the carbon anode has been baked in an oven prior to use in the electrolytic process

Pusher machine operator - a worker who controls the pusher machinery which is located on the side of a coke-oven battery. The pusher machine has a bar that levels the coal charge in the oven door removing apparatus, and a ram for pushing the coke out of the oven

Quench car operator - a worker who drives the quench car, which contains hot coke, to the quenching (cooling) tower and then deposits the cooled coke onto the coke wharf

Sandslinger - a mechanical device used in iron and steel founding for dispersing sand into the moulding box; it may incorporate transportation of the sand from the storage hopper and mixing of the sand with a binder

Shake-out - a process in which a casting is removed from the sand mould, using a vibratory grid or other mechanical or manual methods

Slot oven - the type of coke oven that is in use today. These tall, narrow ovens are arranged in rows to form a battery. The volatiles are collected

Söderberg pot - an electrolytic cell for primary aluminium production in which the anode consists of a mixture of coke and pitch which is baked *in situ* by passage of an electric current and heat from the electrolyte. There are two types: vertical stud and horizontal stud. In the vertical-stud type, the studs through which current enters the anode are set vertically in the top of the anode. In the horizontal-stud type, the studs through which the current enters the anode are set into the sides of the anode

Stud (pin) - a headless metal bolt that is inserted into the anodes of an electrolytic cell for primary aluminium production and that serves to carry the electric current into the anode

Stud puller - a person involved in the removal of studs from a Söderberg pot (normally using a pneumatic press)

Stud setter - a person who inserts studs (pins) into the anode in a Söderberg pot for primary aluminium production

Tapping crew - persons who carry out the siphoning of molten aluminium from the pot into a crucible

Tar chaser - a worker who loosens tar accumulations by means of a rod with a spoon-type end in the collecting main

Utility man - a worker on a coke-oven battery who substitutes for the regular operators during their breaks; also termed a spellsman or reliefman

Venturi scrubber - a gas-cleaning device, in which water is injected into a stream of dust-laden gas flowing at high velocity, thus transferring the dust particles to the water droplets, which are subsequently removed

SUPPLEMENTARY CORRIGENDA TO VOLUMES 1-33

Corrigenda covering volumes 1-6 appeared in volume 7; others appeared in volumes 8, 10-13 and 15-33.

Volume 10

p. 32	Section 2.2, first line	*replace* fungi *by* actinomycetes

Volume 26

p. 349	Molecular formula	*replace* $C_{46}H_{59}N_4O_9.H_2SO_4$ *by* $C_{46}H_{58}N_4O_9.H_2SO_4$

Volume 31

p. 199	Section (*b*), third line	replace mycotoxin A *by* ochratoxin A

Volume 32

p. 291	Line 20	*replace* 27/28 *by* 27% of 28

CUMULATIVE INDEX TO IARC MONOGRAPHS ON THE EVALUATION OF THE CARCINOGENIC RISK OF CHEMICALS TO HUMANS

Numbers in italics indicate volume, and other numbers indicate page. References to corrigenda are given in parentheses. Compounds marked with an asterisk(*) were considered by the working groups in the year indicated, but monographs were not prepared becaused adequate data on carcinogenicity were not available.

IARC SCIENTIFIC PUBLICATIONS

Available from Oxford University Press, Walton Street, Oxford OX2 6DP, UK and in London, New York, Toronto, Delhi, Bombay, Calcutta, Madras, Karachi, Kuala Lumpur, Singapore, Hong Kong, Tokyo, Nairobi, Dar es Salaam, Cape Town, Melbourne, Auckland and associated companies in Beirut, Berlin, Ibadan, Mexico City, Nicosia.

Liver Cancer	No. 1, 1971; 176 pages US$ 10.000; Sw. fr. 30.—
Oncogenesis and Herpesviruses	No. 2, 1972; 515 pages US$ 25.00; Sw. fr. 100.—
N-Nitroso Compounds, Analysis and Formation	No. 3, 1972; 140 pages US$ 6.25; Sw. fr. 25.—
Transplacental Carcinogenesis	No. 4, 1973; 181 pages US$ 12.00; Sw. fr. 40.—
Pathology of Tumours in Laboratory Animals—Volume I—Tumours of the Rat, Part 1	No. 5, 1973; 214 pages US$ 15.00; Sw. fr. 50.—
Pathology of Tumours in Laboratory Animals—Volume I—Tumours of the Rat, Part 2	No. 6, 1976; 319 pages US$ 35.00; Sw. fr. 90.— (OUT OF PRINT)
Host Environment Interactions in the Etiology of Cancer in Man	No. 7, 1973; 464 pages US$ 40.00; Sw. fr. 100.—
Biological Effects of Asbestos	No. 8, 1973; 346 pages US$ 32.00; Sw. fr. 80.—
N-Nitroso Compounds in the Environment	No. 9, 1974; 243 pages US$ 20.00; Sw. fr. 50.—
Chemical Carcinogenesis Essays	No. 10, 1974; 230 pages US$ 20.00; Sw. fr. 50.—
Oncogenesis and Herpesviruses II	No. 11, 1975; Part 1, 511 pages US$ 38.00; Sw. fr. 100.— Part 2, 403 pages US$ 30.00; Sw. fr. 80.—
Screening Tests in Chemical Carcinogenesis	No. 12, 1976; 666 pages US$ 48.00; Sw. fr. 120.—
Environmental Pollution and Carcinogenic Risks	No. 13, 1976; 454 pages US$ 20.00; Sw. fr. 50.—
Environmental *N*-Nitroso Compounds–Analysis and Formation	No. 14, 1976; 512 pages US$ 45.00; Sw. fr. 110.—
Cancer Incidence in Five Continents—Volume III	No. 15, 1976; 584 pages US$ 40.00; Sw. fr. 100.—
Air Pollution and Cancer in Man	No. 16, 1977; 331 pages US$ 35.00; Sw. fr. 90.—
Directory of On-Going Research in Cancer Epidemiology 1977	No. 17, 1977; 599 pages US$ 10.00; Sw. fr. 25.— (OUT OF PRINT)
Environmental Carcinogens—Selected Methods of Analysis, Vol. 1: Analysis of Volatile Nitrosamines in Food	No. 18, 1978; 212 pages US$ 45.00; Sw. fr. 90.—
Environmental Aspects of *N*-Nitroso Compounds	No. 19, 1978; 566 pages US$ 50.00; Sw. fr. 100.—
Nasopharyngeal Carcinoma: Etiology and Control	No. 20, 1978; 610 pages US$ 60.00; Sw. fr. 100.—
Cancer Registration and Its Techniques	No. 21, 1978; 235 pages US$ 25.00; Sw. fr. 40.—
Environmental Carcinogens—Selected Methods of Analysis, Vol. 2: Methods for the Measurement of Vinyl Chloride in Poly(vinyl chloride), Air, Water and Foodstuffs	No. 22, 1978; 142 pages US$ 45.00; Sw. fr. 75.—
Pathology of Tumours in Laboratory Animals—Volume II—Tumours of the Mouse	No. 23, 1979; 669 pages US$ 60.00; Sw. fr. 100.—
Oncogenesis and Herpesviruses III	No. 24, 1978; Part 1, 580 pages US$ 30.00; Sw. fr. 50.— Part 2, 522 pages US$ 30.00; Sw. fr. 50.—
Carcinogenic Risks—Strategies for Intervention	No. 25, 1979; 283 pages US$ 30.00; Sw. fr. 50.—
Directory of On-Going Research in Cancer Epidemiology 1978	No. 26, 1978; 550 pages Sw. fr. 30.—
Molecular and Cellular Aspects of Carcinogen Screening Tests	No. 27, 1980; 371 pages US$ 40.00; Sw. fr. 60.—
Directory of On-Going Research in Cancer Epidemiology 1979	No. 28, 1979; 672 pages Sw. fr. 30.— (OUT OF PRINT)
Environmental Carcinogens—Selected Methods of Analysis, Vol. 3: Analysis of Polycyclic Aromatic Hydrocarbons in Environmental Samples	No. 29, 1979; 240 pages US$ 30.00; Sw. fr. 50.—
Biological Effects of Mineral Fibres	No. 30, 1980; Volume 1, 494 pages US$ 35.00; Sw. fr. 60.— Volume 2, 513 pages US$ 35.00; Sw. fr. 60.—
N-Nitroso Compounds: Analysis, Formation and Occurrence	No. 31, 1980; 841 pages US$ 40.00; Sw. fr. 70.—
Statistical Methods in Cancer Research, Vol. 1: The Analysis of Case-Control Studies	No. 32, 1980; 338 pages US$ 30.00; Sw. fr. 50.—
Handling Chemical Carcinogens in the Laboratory—Problems of Safety	No. 33, 1979; 32 pages US$ 8.00; Sw. fr. 12.—
Pathology of Tumours in Laboratory Animals—Volume III—Tumours of the Hamster	No. 34, 1982; 461 pages US$ 40.00; Sw. fr. 80.—
Directory of On-Going Research in Cancer Epidemiology 1980	No. 35, 1980; 660 pages Sw. fr. 35.—
Cancer Mortality by Occupation and Social Class 1851-1971	No. 36, 1982; 253 pages US$ 30.00; Sw. fr. 60.—
Laboratory Decontamination and Destruction of Aflatoxins B_1, B_2, G_1, G_2 in Laboratory Wastes	No. 37, 1980; 59 pages US$ 10.00; Sw. fr. 18.—
Directory of On-Going Research in Cancer Epidemiology 1981	No. 38, 1981; 696 pages Sw. fr. 40.—
Host Factors in Human Carcinogenesis	No. 39, 1982; 583 pages US$ 50.00; Sw. fr. 100.—
Environmental Carcinogens—Selected Methods of Analysis, Vol. 4: Some Aromatic Amines and Azo Dyes in the General and Industrial Environment	No. 40, 1981; 347 pages US$ 30.00; Sw. fr. 60.—
N-Nitroso Compounds: Occurrence and Biological Effects	No. 41, 1982; 755 pages US$ 55.00; Sw. fr. 110.—
Cancer Incidence in Five Continents—Volume IV	No. 42, 1982; 811 pages US$ 50.00; Sw. fr. 100.—
Laboratory Decontamination and Destruction of Carcinogens in Laboratory Wastes: Some *N*-Nitrosamines	No. 43, 1982; 73 pages US$ 10.00; Sw. fr. 18.—

Environmental Carcinogens—Selected Methods of Analysis, Vol. 5: Mycotoxins	No. 44, 1983; 455 pages US$ 30.00; Sw. fr. 60.—
Environmental Carcinogens—Selected Methods of Analysis, Vol. 6: *N*-Nitroso Compounds	No. 45, 1983 ; 508 pages US $ 40.00; Sw.fr. 80.—
Directory of On-Going Research in Cancer Epidemiology 1982	No. 46, 1982; 722 pages Sw. fr. 40.—
Cancer Incidence in Singapore	No. 47, 1982; 174 pages US$ 15.00; Sw. fr. 30.—
Cancer Incidence in the USSR Second Revised Edition	No. 48, 1982; 75 pages US$ 15.00; Sw. fr. 30.—
Laboratory Decontamination and Destruction of Carcinogens in Laboratory Wastes: Some Polycyclic Aromatic Hydrocarbons	No. 49, 1983; 81 pages US$ 10.00; Sw. fr. 20.-
Directory of On-Going Research in Cancer Epidemiology 1983	No. 50, 1983; 740 pages Sw. fr. 50.-
Modulators of Experimental Carcinogenesis	No. 51, 1983; 307 pages US$ 40.00; Sw. fr. 80.–
Second Cancers Following Radiation Treatment for Cancer of the Uterine Cervix: The Results of a Cancer Registry Collaborative Study	No. 52, 1984; 207 pages US $ 25.00; Sw. fr. 50.–
Nickel in the Human Environment	No. 53, 1984 (in press)
Laboratory Decontamination and Destruction of Carcinogens in Laboratory Wastes: Some Hydrazines	No. 54; 1983; 87 pages US$ 10.00; Sw. fr. 20.–
Laboratory Decontamination and Destruction of Carcinogens in Laboratory Wastes: Some *N*-Nitrosamides	No. 55, 1984; 65 pages US$ 10.00; Sw. fr. 20.-

NON-SERIAL PUBLICATIONS

Available from WHO Sales Agents

Alcool et Cancer	1978; 42 pages Fr. fr. 35-; Sw. fr. 14.-
Information Bulletin on the Survey of Chemicals Being Tested for Carcinogenicity No. 8	1979, 604 pages US$ 20.00; Sw.fr. 40.-
Cancer Morbidity and Causes of Death Among Danish Brewery Workers	1980, 145 pages US$ 25.00; Sw.fr. 45.-
Information Bulletin on the Survey of Chemicals Being Tested for Carcinogenicity No. 9	1981, 294 pages US$ 20.00; Sw.fr. 41.-
Information Bulletin on the Survey of Chemicals Being Tested for Carcinogenicity No. 10	1982, 326 pages US$ 20.00; Sw.fr. 42.-

IARC MONOGRAPHS ON THE EVALUATION OF THE CARCINOGENIC RISK OF CHEMICALS TO HUMANS

Available from WHO Sales Agents. See addresses on back cover.

Some Inorganic Substances, Chlorinated Hydrocarbons, Aromatic Amines, *N*-Nitroso Compounds, and Natural Products	Volume 1, 1972; 184 pages (out of print)
Some Inorganic and Organometallic Compounds	Volume 2, 1973; 181 pages US$ 3.60; Sw. fr. 12.-- (out of print)
Certain Polycyclic Aromatic Hydrocarbons and Heterocyclic Compounds	Volume 3, 1973; 271 pages (out of print)
Some Aromatic Amines, Hydrazine and Related Substances, *N*-Nitroso Compounds and Miscellaneous Alkylating Agents	Volume 4, 1974; 286 pages US$ 7.20; Sw. fr. 18.--
Some Organochlorine Pesticides	Volume 5, 1974; 241 pages US$ 7.20; Sw. fr. 18.-- (out of print)
Sex Hormones	Volume 6, 1974; 243 pages US$ 7.20; Sw. fr. 18.--
Some Anti-thyroid and Related Substances, Nitrofurans and Industrial Chemicals	Volume 7, 1974; 326 pages US$ 12.80; Sw. fr. 32.--
Some Aromatic Azo Compounds	Volume 8, 1975; 357 pages US$ 14.40; Sw. fr. 36.--
Some Aziridines, *N*-, *S*- and *O*-Mustards and Selenium	Volume 9, 1975; 268 pages US$ 10.80; Sw. fr. 27.--
Some Naturally Occurring Substances	Volume 10, 1976; 353 pages US$ 15.00; Sw. fr. 38.--
Cadmium, Nickel, Some Epoxides, Miscellaneous Industrial Chemicals and General Considerations on Volatile Anaesthetics	Volume 11, 1976; 306 pages US$ 14.00; Sw. fr. 34.--
Some Carbamates, Thiocarbamates and Carbazides	Volume 12, 1976; 282 pages US$ 14.00; Sw. fr. 34.--
Some Miscellaneous Pharmaceutical Substances	Volume 13, 1977; 255 pages US$ 12.00; Sw. fr. 30.--
Asbestos	Volume 14, 1977; 106 pages US$ 6.00; Sw. fr. 14.--
Some Fumigants, the Herbicides 2,4-D and 2,4,5-T, Chlorinated Dibenzodioxins and Miscellaneous Industrial Chemicals	Volume 15, 1977; 354 pages US$ 20.00; Sw. fr. 50.--
Some Aromatic Amines and Related Nitro Compounds - Hair Dyes, Colouring Agents and Miscellaneous Industrial Chemicals	Volume 16, 1978; 400 pages US$ 20.00; Sw. fr. 50.--
Some *N*-Nitroso Compounds	Volume 17, 1978; 365 pages US$ 25.00; Sw. fr. 50.--
Polychlorinated Biphenyls and Polybrominated Biphenyls	Volume 18, 1978; 140 pages US$ 13.00; Sw. fr. 20.--
Some Monomers, Plastics and Synthetic Elastomers, and Acrolein	Volume 19, 1979; 513 pages US$ 35.00; Sw. fr. 60.--
Some Halogenated Hydrocarbons	Volume 20, 1979; 609 pages US$ 35.00; Sw. fr. 60.--
Sex Hormones (II)	Volume 21, 1979; 583 pages US$ 35.00; Sw. fr. 60.--
Some Non-nutritive Sweetening Agents	Volume 22, 1980; 208 pages US$ 15.00; Sw. fr. 25.--
Some Metals and Metallic Compounds	Volume 23, 1980; 438 pages US$ 30.00; Sw. fr. 50.--
Some Pharmaceutical Drugs	Volume 24, 1980; 337 pages US$ 25.00; Sw. fr. 40.--
Wood, Leather and Some Associated Industries	Volume 25, 1980; 412 pages US$ 30.00; Sw. fr. 60.--
Some Anticancer and Immunosuppressive Drugs	Volume 26, 1981; 411 pages US$ 30.00; Sw. fr. 62.--
Some Aromatic Amines, Anthraquinones and Nitroso Compounds and Inorganic Fluorides Used in Drinking-Water and Dental Preparations	Volume 27, 1982; 341 pages US$ 25.00; Sw. fr. 40.--
The Rubber Industry	Volume 28, 1982; 486 pages US$ 35.00; Sw. fr. 70.--
Some Industrial Chemicals and Dyestuffs	Volume 29, 1982; 416 pages US$ 30.00; Sw. fr. 60.--
Miscellaneous Pesticides	Volume 30, 1983; 424 pages US$ 30.00; Sw. fr. 60.--
Some Feed Additives, Food Additives and Naturally Occurring Substances	Volume 31, 1983; 314 pages US$ 30.00; Sw. fr. 60. --
Chemicals and Industrial Processes Associated with Cancer in Humans (IARC Monographs 1-20)	Supplement 1, 1979; 71 page (out of print)
Long-term and Short-term Screening Assays for Carcinogens: A Critical Appraisal	Supplement 2, 1980; 426 pa US$ 25.00; Sw. fr. 40.--
Cross Index of Synonyms and Trade Names in Volumes 1 to 26	Supplement 3, 1982; 199 pa US$ 30.00; Sw. fr. 60.--
Chemicals, Industrial Processes and Industries Associated with Cancer in Humans (IARC Monographs Volumes 1 to 29)	Supplement 4, 1982; 292 pa US$ 30.00; Sw. fr. 60.--
Polynuclear Aromatic Compounds, Part 1, Chemical, Environmental and Experimental Data	Volume 32, 1983 ; 477 pages US $ 35.00 ; Sw.fr. 70
Polynuclear Aromatic Compounds, Part 2, Carbon Blacks, Mineral Oils and Some Nitroarenes	Volume 33, 1984; 245 pages US$ 25.00; Sw. fr. 50.--
Polynuclear Aromatic Compounds, Part 3, Industrial Exposures in Aluminium Production, Coal Gasification, Coke Production, and Iron and Steel Founding	Volume 34, 1984; 219 pages US$ 20.00; Sw.fr. 48.--

Composition, impression et façonnage
Groupe MCP-Mame
Dépôt légal : Octobre 1984